Active Ageing and Labour Market Engagement

Zakir Husain

Active Ageing and Labour Market Engagement

Evidence from Eastern India

Zakir Husain
Department of Economics
Presidency University
Kolkata, West Bengal, India

ISBN 978-981-15-0585-0   ISBN 978-981-15-0583-6 (eBook)
https://doi.org/10.1007/978-981-15-0583-6

© Springer Nature Singapore Pte Ltd. 2020
This work is subject to copyright. All rights are reserved by the Publisher, whether the whole or part of the material is concerned, specifically the rights of translation, reprinting, reuse of illustrations, recitation, broadcasting, reproduction on microfilms or in any other physical way, and transmission or information storage and retrieval, electronic adaptation, computer software, or by similar or dissimilar methodology now known or hereafter developed.
The use of general descriptive names, registered names, trademarks, service marks, etc. in this publication does not imply, even in the absence of a specific statement, that such names are exempt from the relevant protective laws and regulations and therefore free for general use.
The publisher, the authors, and the editors are safe to assume that the advice and information in this book are believed to be true and accurate at the date of publication. Neither the publisher nor the authors or the editors give a warranty, expressed or implied, with respect to the material contained herein or for any errors or omissions that may have been made. The publisher remains neutral with regard to jurisdictional claims in published maps and institutional affiliations.

This Springer imprint is published by the registered company Springer Nature Singapore Pte Ltd.
The registered company address is: 152 Beach Road, #21-01/04 Gateway East, Singapore 189721, Singapore

To
My father
For bequeathing a wonderful treasure—a love of books

Acknowledgements

The study was funded by a grant from the Indian Council of Social Science Research, New Delhi, under its Major Research Project scheme. I am grateful to the ICSSR for providing me with the financial assistance to undertake the primary survey and obtain the necessary reference materials and other necessities to complete the project.

During the execution of the project, I received support from various sources that I gratefully acknowledge.

The following persons assisted in the execution of the survey: Ms. Debaki Sarkar and Mr. Roshan Ronghang (Field Supervisors); Dr. Kushal Kumar Sahu, Ms. Mousumi Samal, Mr. Navin Sahu, Ms. Payel Banerjee, Mr. Pravat Kumar Gupta, Mr. Ravi Kumar, Mr. Sourabh Dhar, and Mr. Subhajit Duttagupta (Field Investigators); and Mr. Sukumar Sarkar (Summer Intern). I would also like to thank Ms. Bijayalaxmi Mohanty and Ms. Poonam Priyadarshini Senapati for their assistance during the field survey at Bhubaneswar. Mr. Prabodh Pande (Summer Intern) prepared the data entry package in CS Pro.

The help extended by Ms. Nayantara Biswas, Research Assistant, was invaluable. She helped to clean and edit the data and also undertook the analysis. Chapter 1 was mainly prepared by her. She also painstakingly prepared the reference list, contents, and lists of tables and figures. The methodological manner in which she preserved her work greatly reduced my labour. Her careful editing rooted out many errors, while her willingness to learn was a source of inspiration. I wish her the best in her future academic career.

In addition, the following persons helped me to execute the work: Dr. Saswata Ghosh, Health and Demography Specialist, Centre for Health Policy, Asian Development Research Institute, Patna, who provided useful suggestions and contacts for the field survey; Dr. Debjani Roy, Head of Postgraduate Department of Geography, Nirmala College, Ranchi; and Dr. Sumantra Ray, Presidency University, who helped in the Ranchi component of the survey.

At the administrative level, Mr. Sumit Biswas, Deputy Registrar, Sponsored Research and Industry Cell helped to smoothen all administrative and financial difficulties. The SRIC staff were super-efficient in finance-related matters of the project. Without their ready assistance, I would not have been able to complete the study.

I am particularly grateful to Dr. Inder Sekhar Yadav. He provided mental support and encouragement at all times and in all matters. This was invaluable to me in my work.

Last, but not the least, I am indebted to my mother and wife for their constant support and encouragement. They bore the brunt of my labour without complaint; while their prodding helped me to reinvigorate myself and jump again into the fray. It is said that "women hold up half the sky"; in my case, they held up an even greater share, for which I will be grateful to them.

March 17, 2019   Zakir Husain

Contents

1	**Prolonging of Life Cycle and Its Implications**		1
	1.1 Definition of Ageing		1
	1.2 Global Trends in Ageing		2
	1.3 Ageing in India		6
		1.3.1 Health Challenge: Rising Burden of NCDs	8
		1.3.2 Economic Challenge: Lack of Economic Security	8
		1.3.3 Social Challenge: Feminization of Aged	9
		1.3.4 Government Response: Successes and Failures	10
	1.4 Regional Variations		11
	1.5 Role of Aged in Indian Society		16
	1.6 Research Problem		17
		1.6.1 Conceptual Framework	18
		1.6.2 Research Questions	19
		1.6.3 Database and Methodology	20
	1.7 Implications		21
	1.8 Structure of Study		21
	References		21
2	**Positive Implications of Ageing for Indian Families: Aged as an Asset**		25
	2.1 Introduction		25
	2.2 Active Ageing		25
	2.3 Work Participation of Elderly: A Brief Literature Review		26
		2.3.1 Workforce Participation Among the Aged in India	26
		2.3.2 Work and Earnings	28
	2.4 Aged in the Labour Market		30
		2.4.1 Workforce Participation	30
		2.4.2 Informal Sector and the Aged	31
		2.4.3 Occupational Pattern of Aged Workers	33
		2.4.4 Earnings	35
		2.4.5 Non-financial Contribution of the Elderly	37

	2.5	Conclusion	38
	References		39
3	**Sample Profile**		41
	3.1	Introduction	41
	3.2	NSSO Data	42
		3.2.1 Construction of Variables	43
		3.2.2 Sample Profile	44
	3.3	Primary Survey	47
4	**Economic Contribution of the Aged: A National Profile**		51
	4.1	Introduction	51
	4.2	Methodology	52
		4.2.1 Functional Specification	55
	4.3	Results and Discussion	56
		4.3.1 Financial Contribution of Elderly to Household Expenditure	56
		4.3.2 Changes in Net Financial Contribution Across Socio-economic Strata	59
		4.3.3 Regional Analysis of Financial Contribution of the Elderly	60
	4.4	Determinants of Financial Contribution of Elderly	65
		4.4.1 Econometric Analysis	66
	4.5	Contribution of the Aged and Poverty Levels	69
		4.5.1 Effects of Contribution of Elderly on Household Poverty	71
	4.6	State-Wise Analysis of the Effects of Contribution on Household Poverty	73
		4.6.1 State-Wise Analysis of the Effects of Contribution of Elderly Workers on Household Poverty in the 55th Round	73
		4.6.2 State-Wise Analysis of the Effects of Contribution of Elderly Workers on Household Poverty in the 68th Round	74
	4.7	Change in Household Poverty	76
		4.7.1 Change in Household Poverty in the 55th Round: Link with Proportion of Elderly and Log of State Domestic Product	76
		4.7.2 Change of Household Poverty in the 68th Round: Link with Proportion of Elderly and Log of State Domestic Product	77
	4.8	State-Wise Analysis of the Effects of Financial Contribution of Elderly Workers on the Intensity of Household Poverty	78

		4.8.1	Analysis of FGT Index in the 55th Round.	79
		4.8.2	Analysis of FGT Index in the 68th Round.	81
		4.8.3	FGT Index at the All-India Level.	83
	4.9	Summing Up		84
	Appendix.			86
	References.			92

5 Economic Contribution of the Aged: A Regional Profile ... 93
- 5.1 Introduction ... 93
- 5.2 Workforce Participation ... 93
- 5.3 Nature of Employment ... 94
- 5.4 Earnings and Gross Contribution to Household ... 94
- 5.5 Determinants of Working, Earnings, and Gross Contribution ... 96
- 5.6 Indirect Economic Contribution of Aged ... 99
  - 5.6.1 Aged and Household Chores ... 100
  - 5.6.2 Imputing Economic Value to Household Chores ... 103
  - 5.6.3 Net Contribution of Aged ... 106
  - 5.6.4 Econometric Analysis ... 109
- 5.7 Summing Up ... 110
- References ... 110

6 Daily Life of the Aged: An Analysis of Time-Use Diaries ... 113
- 6.1 What Are Time-Use Studies? ... 113
- 6.2 Studies on Time Use ... 115
- 6.3 Objective and Methodology ... 116
- 6.4 Main Findings ... 119
  - 6.4.1 Mean Time Spent on Activities ... 119
  - 6.4.2 Percentage Time Allocated on Broad Activity Groups ... 120
  - 6.4.3 Variations in Percentage Time Allocated on Broad Activity Groups Across Correlates ... 121
- 6.5 Econometric Analysis ... 123
  - 6.5.1 Active Ageing as Dependent Variable: Ordinary Least Squares Model ... 123
  - 6.5.2 An Alternative Econometric Model ... 124
- 6.6 Conclusion ... 126
- References ... 127

7 Conclusion ... 129
- 7.1 Returning to Research Questions ... 129
- 7.2 Main Findings ... 129
  - 7.2.1 Economic Contribution of the Elderly ... 129
  - 7.2.2 Economic Contribution of the Elderly: A Regional Profile ... 132

		7.2.3	Indirect Contribution of the Elderly:	
			A Regional Profile	133
		7.2.4	Gross Economic Contribution of the Elderly	134
		7.2.5	Net Economic Contribution of the Elderly	135
		7.2.6	Time-Use Pattern of the Aged	136
		7.2.7	Active Ageing in India	136
	7.3	Asset or Burden?		137
	7.4	Some Policy Issues		137
		7.4.1	Promoting Workforce Participation of the Aged	138
		7.4.2	Other Means of Promoting Active Ageing	140
		7.4.3	Health Ageing	142
References				144

About the Author

Zakir Husain is Professor at the Economics Department in Presidency University, Kolkata. He uses econometric tools and methods to study issues related to exclusion and discrimination in gerontology, education, demography, and health. He has been a faculty in Institutes like Indian Institute of Economic Growth, New Delhi, and Indian Institute of Technology, Kharagpur. He has also been a Consultant in the Prime Minister's High Level Committee for preparing a report on Status of Muslim Community in India and a member of the West Bengal State Planning Board.

List of Figures

Fig. 1.1	Population aged 60–79 years and aged 80 years or over by development groups	3
Fig. 1.2	Trend of aged population by income groups	5
Fig. 1.3	Growth of elderly population in India by gender	7
Fig. 1.4	Main workers in Indian labour force by sex and age, 2011	9
Fig. 1.5	Trends in male–female differences in life expectancy at ages 60 and 80 in India, 1950–2055	10
Fig. 1.6	Percentage of elderly population across states in India, 2011	14
Fig. 1.7	Projected percentage of elderly population across states in India, 2026	15
Fig. 1.8	Consumption and income trends as hypothesized in Ando-Modigliani life cycle hypothesis	18
Fig. 1.9	Consumption and income trends in an ageing society	19
Fig. 2.1	Informal sector participation of different age groups in India in the 55th and the 68th rounds of NSS	31
Fig. 2.2	Daily average real earnings of the elderly in the 55th and 68th rounds of NSS	36
Fig. 3.1	Percentage of aged people in population by place of residence and sex	45
Fig. 3.2	Number of respondents by gender and place of residence	48
Fig. 4.1	Percentage of households where elderly are burden or asset in India in the 55th and 68th rounds of NSS	57
Fig. 4.2	Distribution of households by net contribution of aged members in rural and urban India in the 55th and 68th rounds of NSS	57
Fig. 4.3	Gross contribution of the elderly to household expenditure in the 55th and 68th rounds of NSS	58
Fig. 4.4	Net contribution of the elderly to household expenditure in the 55th and 68th rounds of NSS	58

Fig. 4.5 Financial contribution of the elderly in different states of rural India in the 55th round of NSS (percentage) 61

Fig. 4.6 Financial contribution of the elderly in different states of urban India in the 55th round of NSS (percentage) 62

Fig. 4.7 State-wise relationship between average financial contribution and proportion of elderly in rural and urban India in the 55th round of NSS (percentage) 63

Fig. 4.8 State-wise relationship between the average financial contribution and log of net state domestic product in rural and urban India in the 55th round of NSS (percentage) 64

Fig. 4.9 Financial contribution of the elderly in different states of rural India in the 68th round of NSS (percentage) 65

Fig. 4.10 Financial contribution of the elderly in different states of urban India in the 68th round of NSS (percentage) 66

Fig. 4.11 State-wise relationship between the average financial contribution and proportion of the elderly in rural and urban India in the 68th round of NSS (percentage) 67

Fig. 4.12 State-wise relationship between the average financial contribution and log of net state domestic product in rural and urban India in the 68th round of NSS (percentage) 68

Fig. 4.13 Increase in head count ratio in different states after excluding income of aged in rural India in the 55th round (percent) 74

Fig. 4.14 Increase in head count ratio in different states after excluding income of the aged in urban India in the 55th round (percentage) 75

Fig. 4.15 Increase in head count ratio in different states after excluding income of the aged in rural India in the 68th round (percentage) 76

Fig. 4.16 Increase in head count ratio in different states after excluding income of the aged in urban India in the 68th round (percentage) 77

Fig. 4.17 Change in poverty in different states across the proportion of elderly in rural and urban India in the 55th round of NSS (percentage) 78

Fig. 4.18 Change in poverty in different states across the net state domestic product in rural and urban India in 55th round of NSS (percentage) 79

Fig. 4.19 Change in poverty in different states across the proportion of elderly in rural and urban India in the 68th round of NSS (percentage) 80

Fig. 4.20 Change in poverty in different states across the net state domestic product in rural and urban India in 68th round of NSS (percentage) 80

Fig. 4.21 Increase in FGT index in different states after excluding income of aged in rural India in the 55th round (percentage) 81

Fig. 4.22 Increase in FGT index in different states after excluding income of the aged in urban India in the 55th round (percentage) 82

Fig. 4.23	Increase in FGT index in different states after excluding income of the aged in rural India in the 68th round (percentage)	83
Fig. 4.24	Increase in FGT index in different states after excluding income of aged in urban India in the 68th round (percentage)	84
Fig. 4.25	FGT index, including and excluding contribution of the aged, by place of residence and rounds	85
Fig. 5.1	Workforce participation by gender across cities	94
Fig. 5.2	Proportion of workers engaged in different occupation	96
Fig. 5.3	Mean number of days in a week that a household chore is performed	102
Fig. 5.4	Mean number of days in a week that a child-related chore is performed	105
Fig. 6.1	Mean time spent on different activities as share of total day	120
Fig. 6.2	Percentage of time spent on broad activity group	121
Fig. 6.3	Kernel density of percentage hours spent actively	125
Fig. 6.4	Histogram of percentage hours spent actively	125
Fig. 7.1	Stylized graph of age, labour productivity, and age	139

List of Tables

Table 1.1	Snapshot view of government policies for the welfare of aged	12
Table 2.1	Occupational segregation between the elderly and near elderly in the 55th and the 68th rounds of NSS	33
Table 3.1	Sample profile of aged persons in 55th round of NSSO	45
Table 3.2	Sample profile of aged persons in 68th round of NSSO	46
Table 3.3	Mean age, household size, and per capita income by gender and place of residence (NSSO, 55th round)	46
Table 3.4	Mean age, household size, and per capita income by gender and place of residence (NSSO, 68th round)	47
Table 3.5	Sample profile of respondents of primary survey	48
Table 3.6	Mean age, household size, and per capita income by gender and city	49
Table 4.1	Net financial contribution of the elderly workers by expenditure group and location of residence in 55th and 68th rounds of NSS (percentage)	59
Table 4.2	Effects of predictor variables on net financial contribution of elderly in rural India in the 55th and 68th rounds of NSS	69
Table 4.3	Effects of predictor variables on net financial contribution of elderly in urban India in the 55th and 68th rounds of NSS	70
Table 4.4	Head count ratio of rural and urban India in the 55th and 68th rounds of NSS	72
Table 4.5	Net contribution of elderly to household expenditure of different states in India in the 55th and 68th rounds of NSS	86
Table 4.6	Estimated 't' value for the contribution of elderly to household expenditure of different states in India in the 55th and 68th rounds of NSS	87
Table 4.7	Proportion of elderly in the population	88
Table 4.8	Head count ratio in rural and urban India in the 55th round of NSS	89
Table 4.9	Head count ratio in rural and urban India in the 68th round of NSS	89

Table 4.10	FGT index in rural and urban India in the 55th round of NSS	90
Table 4.11	FGT index in rural and urban India in the 68th round of NSS	91
Table 5.1	Workforce participation by socio-demographic correlates	95
Table 5.2	Occupational distribution of aged workers by gender and place of residence	96
Table 5.3	Mean of gross contribution by aged to households	97
Table 5.4	Results of logit model for determinants of working	98
Table 5.5	Heckman model for determinants of working and earnings	99
Table 5.6	Results of models for gross contribution	100
Table 5.7	Frequency of performing household tasks by aged	101
Table 5.8	Frequency of performing child-related chores by aged	104
Table 5.9	Economic valuation of contribution of aged and expenditure on aged	107
Table 5.10	Average percentage net contribution of aged by correlates—by gender	108
Table 5.11	Results of OLS model for net contribution of aged	109
Table 6.1	Activities recorded and their classification	118
Table 6.2	Mean time spent on each activity (minutes)	119
Table 6.3	Variations in proportion of time allocated to broad activity group	122
Table 6.4	Results of OLS model	124
Table 6.5	Results of Tobit model	126

List of Tables

Table 1.1	Snapshot view of government policies for the welfare of aged	12
Table 2.1	Occupational segregation between the elderly and near elderly in the 55th and the 68th rounds of NSS	33
Table 3.1	Sample profile of aged persons in 55th round of NSSO	45
Table 3.2	Sample profile of aged persons in 68th round of NSSO	46
Table 3.3	Mean age, household size, and per capita income by gender and place of residence (NSSO, 55th round)	46
Table 3.4	Mean age, household size, and per capita income by gender and place of residence (NSSO, 68th round)	47
Table 3.5	Sample profile of respondents of primary survey	48
Table 3.6	Mean age, household size, and per capita income by gender and city	49
Table 4.1	Net financial contribution of the elderly workers by expenditure group and location of residence in 55th and 68th rounds of NSS (percentage)	59
Table 4.2	Effects of predictor variables on net financial contribution of elderly in rural India in the 55th and 68th rounds of NSS	69
Table 4.3	Effects of predictor variables on net financial contribution of elderly in urban India in the 55th and 68th rounds of NSS	70
Table 4.4	Head count ratio of rural and urban India in the 55th and 68th rounds of NSS	72
Table 4.5	Net contribution of elderly to household expenditure of different states in India in the 55th and 68th rounds of NSS	86
Table 4.6	Estimated 't' value for the contribution of elderly to household expenditure of different states in India in the 55th and 68th rounds of NSS	87
Table 4.7	Proportion of elderly in the population	88
Table 4.8	Head count ratio in rural and urban India in the 55th round of NSS	89
Table 4.9	Head count ratio in rural and urban India in the 68th round of NSS	89

Table 4.10	FGT index in rural and urban India in the 55th round of NSS	90
Table 4.11	FGT index in rural and urban India in the 68th round of NSS	91
Table 5.1	Workforce participation by socio-demographic correlates	95
Table 5.2	Occupational distribution of aged workers by gender and place of residence	96
Table 5.3	Mean of gross contribution by aged to households	97
Table 5.4	Results of logit model for determinants of working	98
Table 5.5	Heckman model for determinants of working and earnings	99
Table 5.6	Results of models for gross contribution	100
Table 5.7	Frequency of performing household tasks by aged	101
Table 5.8	Frequency of performing child-related chores by aged	104
Table 5.9	Economic valuation of contribution of aged and expenditure on aged	107
Table 5.10	Average percentage net contribution of aged by correlates—by gender	108
Table 5.11	Results of OLS model for net contribution of aged	109
Table 6.1	Activities recorded and their classification	118
Table 6.2	Mean time spent on each activity (minutes)	119
Table 6.3	Variations in proportion of time allocated to broad activity group	122
Table 6.4	Results of OLS model	124
Table 6.5	Results of Tobit model	126

Chapter 1
Prolonging of Life Cycle and Its Implications

1.1 Definition of Ageing

Ageing is rapidly becoming an important issue in both developed and developing countries. It comprises a widely discussed topic among researchers, policy makers, media, and even the common people. But what is ageing? Ageing is defined in dictionaries as 'the process of growing old' (Hornby et al. 2010). An online source (https://www.medicinenet.com/script/main/art.asp?articlekey=13403 accessed on 4 February 2018) defines ageing as 'the process of becoming older, a process that is genetically determined and environmentally modulated'. Operationally, an aged person is defined as someone above 60 years (in some countries, above 65 years) of age.

But, perhaps, it is better to contemplate ageing as a process. Evolutionary biologists have defined ageing as a 'persistent decline in the age-specific fitness components of an organism due to internal physiological degeneration' (Rose et al. 2012). This approach conceptualizes ageing as a loss: 'aging is defined as a decline or loss (a "de-tuning") of adaptation with increasing age' (Flatt 2012: 1). At the level of the individual, ageing is manifested by the intrinsic physiological state of the organism at its current age; important indications are, among other things, whether the individual is dead or alive and how much she/he reproduces. At the level of the cohort, this translates into age-specific rates of mortality and reproduction.

In contrast, sociologists view the process of ageing as occurring over the life cycle, resulting in maturation and changes observed at the physical, psychological, and social levels (Riley 1978). In the biological level, for instance, ageing is manifested in changes like reduced activity, onset of mobility-related problems, appearance of wrinkles, loss of hair, etc. Some of the changes are due to molecular and cellular changes; they are called **primary ageing**. On the other hand, some of the biological changes are due to controllable factors such as lack of physical exercise and poor diet, and are called **secondary ageing** (Whitbourne and Whitbourne 2010). Socially, ageing is associated with a withdrawal from active roles played within the family and community. While retirement from work is considered an important form of diminishing responsibility, there may be other forms of withdrawal—like physical or mental alienation from family members, lack of participation in decision-making, undertaking household duties, etc. Recent studies, however, argue that such withdrawal may be compensated by involvement in new hobbies or activities. Finally, ageing is also a process of adjusting to mortality and death. Beginning with Kubler-Ross's classic work on dying (Kübler-Ross 1969), this has become an important issue for research in recent years (Husain and Dutta 2015; Agarwal et al. 2016).

This approach generates an important concept, successful or active ageing. WHO defines **healthy ageing** 'as the process of developing and maintaining the **functional ability** that enables wellbeing in older age' (WHO 2015). Functional ability is about having the capabilities that enable all people to be and do what they have reason to value. This includes a person's ability to:

- meet their basic needs
- to learn, grow, and make decisions
- to be mobile
- to build and maintain relationships
- to contribute to society

Functional ability is made up of the intrinsic capacity of the individual, relevant environmental characteristics, and the interaction between them. Intrinsic capacity comprises all the mental and physical capacities that a person can draw on and includes their ability to walk, think, see, hear, and remember. The level of intrinsic capacity is influenced by a number of factors, such as the presence of diseases, injuries, and age-related changes. Environments include the home, community and broader society, and all the factors within them such as the built environment, people and their relationships, attitudes and values, health and social policies, the systems that support them, and the services that they implement. Being able to live in environments that support and maintain your intrinsic capacity and functional ability is key to *Healthy Ageing* (WHO 2015).

1.2 Global Trends in Ageing

The global population is growing older. According to data from *World Population Ageing Report* (UN 2015), the number of people above the age of 60 has accelerated in recent years, in most parts of the world, and that growth is projected to show an upward trajectory in the recent future. Two-thirds of the world's aged population live in the developing countries and their numbers are growing more rapidly there than in the developed nations.

Figure 1.1 displays the rise in the absolute numbers of the elderly population, over the years, by age. The rate of growth of the aged in the more developed regions is projected to have a smaller impact on the increasing number of the elderly in the world population. While the number of people aged 60 years or over in developed

1.2 Global Trends in Ageing

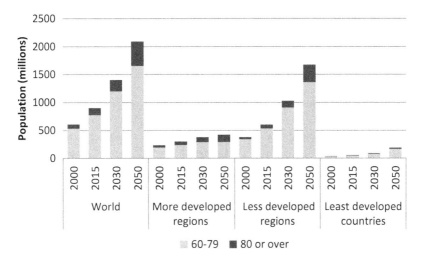

Fig. 1.1 Population aged 60–79 years and aged 80 years or over by development groups. (Source: United Nations (2015). World Population Prospects: The 2015 Revision. (UN 2015))

regions grew from 231 million to 299 million between 2000 and 2015, it is projected to attain 375 million in 2030, thus implying a fall in the growth rate of the aged population by 3%.

In contrast, in the developing regions, the growth rate of the population over the age of 60 years is rising rapidly. The number of older persons in the less-developed regions has grown by 60% in just 15 years, between 2000 and 2015, and it is projected to increase further by 71% in the next 15 years, between 2015 and 2030. By 2050, it is projected that nearly 80% of the world's aged population, a total of 1.7 billion people, will reside in the less-developed regions.

While the absolute numbers of the aged is the lowest in the least developed countries, there has been an increase in the growth rates of the population aged 60 years or over by 54%, between 2000 and 2015. Despite such swift growth, the least developed countries are projected to account for just 6.3% of the world's aged population in 2030 and 8.9% in 2050, as opposed to 5.8% in 2015. In 2015, the least developed countries were home to 3.8% of the global population aged 80 years or over (oldest-old population), and by 2050 their share of the world's oldest-old persons is projected to rise to 4.8%.

Another important demographic trend arising from the above factors is the increase in 80 years and above population:

> As a result of both improved longevity and the ageing of large cohorts (that is, the "baby boomers" born during the post-World War II period), the world's older population is projected to become increasingly aged. Globally, the share of the older population that is aged 80 years or over rose from 9 per cent in 1980 to 14 per cent in 2015 (figure II.8), and it is projected to remain fairly stable between 2015 and 2030. Between 2030 and 2050 the proportion of the world's older persons that are aged 80 years or over is projected to rise from 14 per cent to more than 20 per cent (UN 2015:20).

The report on *Global Health and Ageing* (Suzman and Beard 2011) identifies three major factors responsible for the projected ageing of the global population:

1. *Declining fertility*: Historically fertility has been the most influential in shaping trends in the numbers and proportion of older persons in the population over the long term. Total fertility rates have fallen in each of the world's regions. While the decline began first in Europe, Northern America, and the developed countries of Oceania in the late nineteenth century, a similar trend followed in Asia, Latin America and the Caribbean, and Africa from the mid-twentieth century. A rise in the proportion of older people, as compared to younger members, in the population leads to a higher absolute number of the aged. It has been observed that in 2017, one in eight people globally were above the age of 60. In 2050, the aged are projected to account for one in five people worldwide.
2. *Increase in life expectancy*: While a lower mortality rate for the elderly causes a higher life expectancy in developed countries in Oceania, Europe, and Northern America, higher survival for younger people has a greater impact on improving life expectancy in the less-developed nations, like Africa, Asia, and Latin America and the Caribbean. On a global scale, the 'oldest old' (persons above the age of 80) comprises 8% of the total aged population.
3. *Decline in death rate*: In the post-Industrial Revolution era, gradual improvements in living standards, especially more nutritious diets and cleaner drinking water, began to reduce serious infections and prevent deaths among children. The immunization programmes in the twentieth century eradicated many infectious diseases such as smallpox, polio, etc. This not only reduced child and infant mortality, but also led to an improvement in adult heath, as most of the non-communicable diseases have their origins in poor child health, or even events before birth (Barker 1980). In the twentieth century, mortality began declining among the older adults and aged, a result of improvements in public health. Consequently, more children were surviving their vulnerable early years, reaching adulthood and surviving after 60 years.

Figure 1.2 shows the decomposition of the growth in the absolute numbers of the elderly population by income groups. Growth in the number of the aged population has been fastest in middle-income countries between 2000 and 2015, and it has been further projected that, between 2015 and 2030, this group of countries would be most likely to observe the fastest growth in the absolute numbers of the older population. In this regard, the role of the upper-middle-income countries must be highlighted. Between 2015 and 2030, there was an increase in the absolute number of elderly persons from 195 million to 320 million people, resulting in a 64% rise in this period. It has been anticipated that this number will increase to 545 million people aged above 60 years in 2030, thus resulting in a 70% growth between 2015 and 2030. At the same time, the older population residing in the lower-middle-income countries is projected to rise from 238 million to 394 million contributing to an increase by 66%.

A moderate rise in the absolute number of the elderly population has been observed in the high-income countries. However, there is expected to be a fall in the

1.2 Global Trends in Ageing

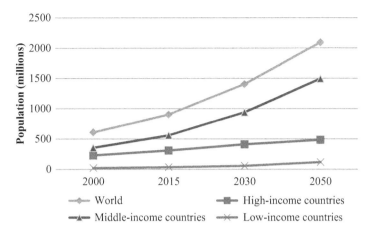

Fig. 1.2 Trend of aged population by income groups. (Source: United Nations (2015). World Population Prospects: The 2015 Revision. (UN 2015))

percentage change of the aged by 2% between 2015 and 2030, compared to 15 years earlier. Due to longer average survival in the high-income countries compared to the other income groups, a large number of the world's oldest-old population are concentrated in these countries. While the proportion is projected to decline somewhat in 2050, it has the highest share in the absolute numbers of the 80 and above population with nearly 61 million people residing in these countries in 2015 which is projected to increase further to almost 91 million persons in 2030.

Although the growth in the population of the aged is rising and is projected to accelerate in the coming years, it is occurring at a slower pace in the low-income countries where the population of the aged grew by 49% and 56%, in 2000 and 2015, respectively. Between 2015 and 2030, the older population in low-income countries is projected to grow by 63%, from 33 million to 54 million. Hence, the low-income countries are expected to see an 81% increase in the number of oldest-old persons between 2015 and 2030. This projection is the most rapid rise observed out of the other income groups. This is expected owing to the fact that the oldest-old persons residing in the low-income countries were born around the middle of the twentieth century in a situation of extremely high fertility.

These trends are going to throw up a host of issues and challenges to society. Four major economic costs that need to be borne due to an older population (World Health Organization 2002; Centre for Ageing Research and Development in Ireland 2011):

4. *Supporting older people in retirement*: An ageing global population means that in a few years, the ratio of people being supported by welfare systems to the people in employment is likely to rise. Due to the fact that average consumption has been exceeding average labour income in ageing societies, pensions are expected to remain the largest expenditure item for countries which provide them. It has been projected that pensions would increase on an average of 3% of

the world's GDP by 2050. Unfortunately, many developing countries have no pension system, or the system is confined to exclusive groups such as military or government officials. Many governments are now recognizing the need to support older people in the workplace by providing incentives to stay in work. These include flexible hours and abolishing the idea of a default retirement age.

5. *Healthcare*: Owing to the shift in disease pattern, non-communicable diseases are projected to contribute 54% of the illnesses plaguing the elderly population in middle- and low-income countries by 2030. It is estimated that this will cause governments of developed countries to raise its healthcare costs by 4.8% of the global GDP by 2050. The developing world is at a serious disadvantage since it is growing old faster than it can grow rich.

6. *Long-term care*: As chronic diseases become a greater burden, it is imperative for countries to adopt solutions to long-term care of disabilities, which may occur due to old age. Improvements in technology are a way to meet the challenge of increasing government spending on healthcare. Another sustainable way is the introduction of 'home care', which can provide the aged the kind of support which they need. This kind of residential healthcare models of handling either acute treatments or chronic disease has gained momentum around the world.

7. *Age-sensitive urban planning:* In order to ensure active ageing, the Framework on Active Ageing has argued that measures should be taken for preventing chronic disease and creating aged-friendly environments (particularly with respect housing, transport, healthcare institutions, financial institutions, and other public places) through sensitive urban planning. This will put the aged in control of their future and enables them to enjoy a more productive life.

1.3 Ageing in India

In Asia, the proportion of the older population is projected to increase from 10.5% to 22.4% between 2012 and 2050. Among the South Asian Association for Regional Cooperation (SAARC) countries, it has been projected that they should attain an average of 21% of its population above 60 years by 2050. While India is not expected to report more than 19% of aged population by 2050, the absolute numbers will still be very large (UN 2017).

India has been undergoing a demographic change moving from a 'young' country to one housing an increasing proportion of the aged. Declining rates of mortality and fertility along with higher life expectancy is responsible for this shift in India's population structure. Figure 1.3 depicts this rise in the elderly population, which has been observed over the past century. In 1951, older persons were almost 20 million. After three decades, it was a little over 43 million, another decade later in 1991, the population had increased to 55.3 million and for 2001 this was estimated at 76 million. The 2011 Census reports that there are 103.9 million aged in India in 2011, of which the majority resides in rural areas (73.3 million in rural areas, and 30.6 mil-

1.3 Ageing in India

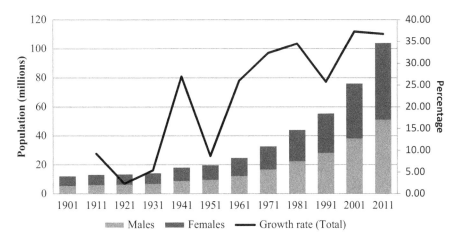

Fig. 1.3 Growth of elderly population in India by gender. (Source: Ageing in India: Occasional Paper No.2 of 1991, Office of the Registrar General & Census Commissioner, India. (Shiksha et al. 2008))

lion in urban areas). There are 51.1 million males, slightly outnumbered by females (52.8 million). As percentage share of population, aged comprise 8.6% in 2011; in rural and urban areas the percentage shares are 8.8 and 8.1%, respectively. Elderly females comprise 9.0% of the female population, while the corresponding figure for males is 8.2%. The increasing rate of growth of the older population is clearly represented by the stacked bar graphs, which shows that the composition of male and female aged persons has roughly remained the same over the past 100 years. These trends exhibit the growing contribution of the older segment in the population of India (Shiksha et al. 2008).

It is interesting to note that the rate of growth of the aged in India is three times higher than that of the population as a whole (Giridhar et al. 2014). India's fertility rate has fallen from 5.9 to 2.3 between 1950 and 2013 and is further projected to drop to 1.88 by 2050, which is below the replacement level of 2.1 children per woman. In addition to this, life expectancy at birth has improved over the past few decades, rising from 36.2 years to 67.5 years between 1950 and 2015 and is projected to increase further to 75.9 years by 2050. Even more significant in its impact on population ageing, life expectancy at age 60 has also increased vastly, increasing from about 12 years to 18 years between 1950 and 2015 and is projected to further increase to more than 21 years by 2050. Similarly, the average Indian life expectancy at age 80 has increased significantly, from about 5 years in 1950 to more than 7 years at the present time. By 2050, it is projected to rise to 8.5 years (UN 2015).

Due to competing development priorities, governments of rapidly ageing developing countries, such as India, are slower in responding to the demographic shift observed by a rising proportion of the elderly in the total population. The changing population structure gives rise to many issues that need to be brought to the attention of policy makers since they are proving to be impediments in the lives of the elderly in India.

1.3.1 Health Challenge: Rising Burden of NCDs

A changing population structure is accompanied by a shift in the health profile, with increasing cases of non-communicable diseases (NCDs), which are more prevalent at the older ages. While such chronic illnesses are on the rise, still infectious diseases alongside violence and injury—especially gender-based violence—continue to plague the elderly in India. These three physical maladies constitute the so-called triple burden of disease in the subcontinent (Bloom et al. 2014). The burden of NCDs has seen substantial growth, between 1990 and 2013, they have surpassed infectious, nutritional, maternal, and perinatal conditions as a cause of death, both in absolute numbers and percentages. Chronic illnesses, mainly cardiovascular illnesses, cancers, and chronic respiratory diseases, have likewise surpassed these other conditions in the number of annual DALY.[1]

There are four major chronic disease risk factors which are fuelling the growth of NCDs in India: tobacco consumption, obesity and physical inactivity (Desai et al. 2010), and alcoholism and mental disorders (Institute for Health Metrics and Evaluation 2013). This growing burden of both physical and mental chronic illnesses has the potential to translate into staggering economic losses; predictions suggest that NCDs may cost India as much as USD 4.3 trillion in productivity losses and healthcare expenditure between 2012 and 2030, a figure that is twice the country's annual GDP (Bloom et al. 2014).

1.3.2 Economic Challenge: Lack of Economic Security

High instances of migration and urbanization have led to a higher percentage of the elderly living alone, especially in the rural areas. Owing to the decline of a family support system there is an absence of a financial safety net for the elderly population in India. Estimates from the census report observe that 42% of the 60+ and 22% of 80+ population are labour force participants (Government of India 2011). As depicted in Fig. 1.4, the main workers in the Indian labour market have shown high levels of work force participation, which are significantly higher for males than for females. It is interesting to note that practically 24% of India's nearly 10,000 aged, who were surveyed by the UNFPA, remained in the workforce and that a more than half of these respondents were working at a relatively high intensity level of at least 6 months out of the year or more than 4 h a day. Furthermore, it has been estimated that more than 70% of aged Indian workers surveyed cited economic necessity, rather than personal preference, as their main reason for remaining in the workforce, indicating a high level of income insecurity (Alam et al. 2012).

---

[1] Disability-adjusted life years (DALYs) are a measure of healthy life used by the World Health Organization and other health-monitoring bodies.

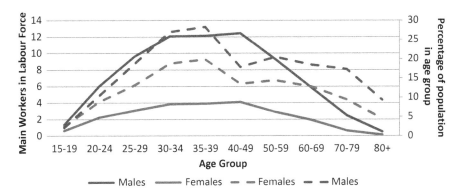

Fig. 1.4 Main workers in Indian labour force by sex and age, 2011. (Source: Estimated from Census of India (Government of India 2011))

An estimated 83% of India's labour force is in the unorganized sector and is not entitled to any pension or retirement benefits in their old age. It has been observed that less than 10% of the total Indian population receives a pension of any kind (Uppal and Sarma 2007). As discussed previously, the prevalence of NCDs increases with age, which leads to a rise in healthcare spending, which proves to be a burden to the elderly with no income security. The average annual per capita out-of-pocket health spending in India is almost four times as high for the elderly as compared to adults below the age of 60 (Bloom et al. 2010). In the absence of adequate health insurance extensive healthcare spending can push the elderly into poverty. This is more notable in rural India where quality healthcare infrastructure is virtually absent or inaccessible.

1.3.3 Social Challenge: Feminization of Aged

Accompanying the changing population structure in India is the higher incidence of feminization in older age groups: the sex ratio of the elderly has increased from 938 women to 1000 men to 1033 women between 1971 and 2011 and is projected to rise to 1060 by 2026. Although there has been a dramatic increase in the average life expectancy in India, it has not risen equally for males and females. As shown in Fig. 1.5, in 1950–1955, Indian women's life expectancy at age 60 exceeded men's by 0.07 years; by 2010–2015, this gap had doubled, and by 2050–2055 it is projected to reach 2 years. Meanwhile, although the male-female gap in life expectancy at age 80 fell between 1950 and the present, it is expected to rise again over the next 40 years (UN 2015).

The rising longevity gap between the sexes implies that India's aged population is growing increasingly female. In 1950, India's population of female adults above the age of 60 was 50.8%. In 2015, despite a high overall male/female sex ratio throughout the latter half of the twentieth century (about 107 males per 100 females),

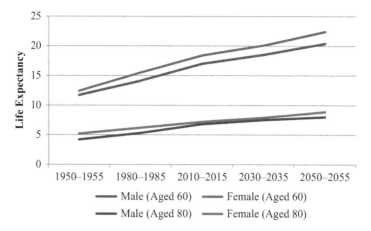

Fig. 1.5 Trends in male–female differences in life expectancy at ages 60 and 80 in India, 1950–2055. (Source: United Nations (2015); 2030–2055 figures are projections based on a medium-fertility scenario. (UN 2015))

this proportion has grown to approximately 52.5% and is projected to reach about 53% by 2050 under a medium-fertility scenario. In the oldest old segment of adults 80 and up, the proportion of females is projected to increase from the current 55% to 56% by 2050. Although the change in percentage points is somewhat small, in absolute terms it represents hundreds of thousands of individuals.

An outcome of this form of a feminization of ageing is the neglect experienced by women as they age, which is often worsened by discrimination due to widowhood. Loss of spouse in old age adds to the vulnerability in later years. A combination of higher life expectancies of females and higher average age of males at first marriage are contributing to a rise in widowed females in India. In 2012, only 8% of Indian males aged 60–64 were widowed compared with 35% of females in this age group. Among adults 80 and older, more than 60% of females had been widowed compared with just 27% of males (Desai et al. 2010). Most notably, widowed females may suffer from income insecurity due to inheritance traditions that favour sons and insecurity in their living arrangements (Dey et al. 2012). Evidence throws light on the fact that female widows aged 60 and above, in India, suffer from morbidity due to prevalence of both communicable illnesses and NCDs at a rate of 13% or higher than do their male counterparts, in the same age group. Work force participation among women is also very low, and a majority of them depend on their families for economic support (Government of India 2011).

1.3.4 Government Response: Successes and Failures

The role of the family as the primary care-giver for the elderly is now changing due to the widespread demographic and socio-economic-cultural changes taking place in these transitional societies. Hence, the government is called upon for the provision of public welfare programmes which will assist the aged in India. This began in 1999 with the National Policy for Older Persons which was undertaken by the Government of India to encourage old age assistance programmes. Presently, the debate on provision of social security to the elderly revolves around the eligibility, coverage, pension amount, delivery mechanisms, suitability, and the economic implications of such measures (Subrahmanya 2017).

Apart for the ones mentioned below, there are a few noteworthy social security benefits provided to workers in the organized sector such as the Employees' Provident Fund Act 1952, Family Pension Scheme 1971, Payment of Gratuity Act 1972, Deposit-linked Insurance Scheme 1976, Group Insurance and General Provident Fund Scheme 1982 and finally, the National Pension Scheme 2004. For workers in the unorganized sector and other weaker sections of population, the Government of India and several state governments have launched subsidized insurance schemes through the Life Insurance Corporation of India and General Insurance Corporation of India via different policies such as Jeevan Akshay for the self-employed, endowment plans, which are savings-linked insurance plans like Jeevan Mitra policy, money back policies that are designed to provide old age security through lump sum benefits over periodic intervals. Table 1.1 presents a summary of welfare scheme for the aged in different sectors of the Indian economy.

1.4 Regional Variations

India has a variety of inter-regional and inter-state demographic diversity based on considerable differences in the population age structure, which is not spread equally among all the states. For instance, as depicted in Fig. 1.6, the southern states, especially Kerala and Tamil Nadu, are the front runners in population ageing along with Himachal Pradesh, Maharashtra, Orissa, and Punjab. The central and northern states such as Bihar, Uttar Pradesh, Madhya Pradesh, and Jharkhand have much lower proportions of aged population.

According to data from the Sample Registration System (SRS), life expectancy of the aged has risen from 14 years to 18 years between the period 1970–1975 and 2010–2014, with women living at an average of 2 years longer than men. All the Indian states show a life expectancy for the aged as over 15 years currently, except males in Chhattisgarh. Thus life expectancy improvement has been substantial in nearly all the states of India, which has led to a rise in mortality of the aged compared to the younger population (Fig. 1.7).

This national trend leaves room for a high degree of heterogeneity across the states. Currently all the Indian states have higher life expectancies at old ages for women than for men. Different male and female life expectancies and life expectancy gaps in different states and regions of the country imply that states will have increasingly dissimilar sex divisions among their older adult populations. Comparing state-level sex differences in life expectancy at age 60 in some of India's most populous states has brought attention to the large range between the half-year difference

Table 1.1 Snapshot view of government policies for the welfare of aged

Sector	Schemes/programmes	Objectives	Year of implementation
Health	National Programme for Health Care of Elderly (NPHCE)	Ensure preventive, curative, and rehabilitative services to the elderly. Strengthen referral system and develop specialized knowledge in the field of diseases related to old age.	2010–2011
	Rashtriya Swasthya Bima Yojana (RSBY)	The beneficiary is any Below Poverty Line (BPL) family, whose information is included in the district BPL list prepared by the state government. They are entitled to hospitalization coverage up to Rs. 30,000 per annum on family floater basis, for most of the diseases that require hospitalization.	2008
Finance and Revenue	Incentives under Income Tax Act, 1961	A senior citizen is liable to pay income tax for income above Rs. 3 lakh per annum and Rs. 5 lakh for people 80 years and older; limits are periodically revised.	1961
	Concessions	Senior citizen concessions in railways—40% for men and 50% for women, Air India offers 50% discount to senior citizens (above 63 years old). Discounted tickets for public road transport.	Varies
Legal/Law	Senior Citizen Savings Scheme	Senior citizens are eligible for tax deduction under section 80C of the Income Tax Act as well as higher interest rates for savings accounts at national banks.	2004
	Maintenance and Welfare of Parents and Senior Citizens	Legally obligates children and heirs to provide maintenance to senior citizens and parents, by monthly allowance, in addition to caring for elderly parents.	2007
Social Justice and Empowerment	Integrated Programme for Older Persons (IPOP)	Providing basic amenities like shelter, food, medical care, and entertainment. Financial assistance is provided to NGOs for maintenance of old-age homes.	1992
	Old-age pension under Indira Gandhi National Old Age Pension Scheme (IGNOPS)	Central government assistance of Rs. 200 per month to people in 60–79 year age group and Rs. 500 to people above 80 years of age belonging to BPL households; supplemented by state governments in varying amounts.	2007
	Annapurna scheme	Providing food security to meet the requirement of those senior citizens who have remained uncovered under the National Old Age Pension Scheme (NOAPS) by giving them 10 kg of food grains per month.	2000–2001

1.4 Regional Variations

Rural Development	Indira Gandhi National Widow Pension Scheme (IGNWPS)	Pension to widows belonging to BPL category of Rs. 200 per month between the ages 40 and 79 and Rs. 500 per month thereafter.	2009
	National Family Benefit Scheme (NFBS)	Central assistance of Rs. 20,000 provided to a BPL household on the death of the primary earner of the family who was below the age of 60.	1998
	Mahatma Gandhi National Rural Employment Guarantee Act (MGNREGA)	The Act guarantees 100 days of employment in a financial year to any rural household whose adult members are willing to do unskilled manual work.	2006
Retirement/ Pension	National Pension System (NPS)	Provides retirement income to all citizens along with market-based returns over long run.	2004
	Swavalamban Scheme	The government contributes a sum of Rs. 1000 for unorganized workers who are NPS subscribers and who contribute a minimum of Rs. 1000 and maximum of Rs. 12,000 per annum.	2010–2011

Source: Population Aging in India: Facts, Issues, and Options Agarwal et al. (2016)

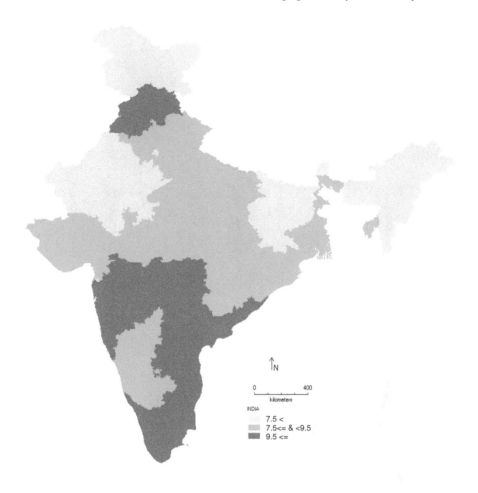

Fig. 1.6 Percentage of elderly population across states in India, 2011. (Source: Estimated by author from Census of India. (Government of India 2011))

in Bihar to the more than four-year gap in Rajasthan, between the life expectancies of females and males. This suggests that different states may see highly different gender profiles in their populations of older adults in the years to come, which might require the state and local governments to tailor their policies and programmes accordingly (Agarwal et al. 2016).

Based on 2011 Census, the overall old-age dependency ratio shows that there are over 14 elderly per 100 working age population, with significant variations across states. In Kerala, Goa, Punjab, Himachal Pradesh, Tamil Nadu, Maharashtra, Orissa, and Andhra Pradesh, the old-age dependency ratio is more than 15% (nearly 20% in Kerala), whereas it is lower than 10% in Arunachal Pradesh, Meghalaya, Nagaland, and Chandigarh. Higher old-age dependency reflects higher level of demand for care from immediate family (UNFPA 2017).

1.4 Regional Variations

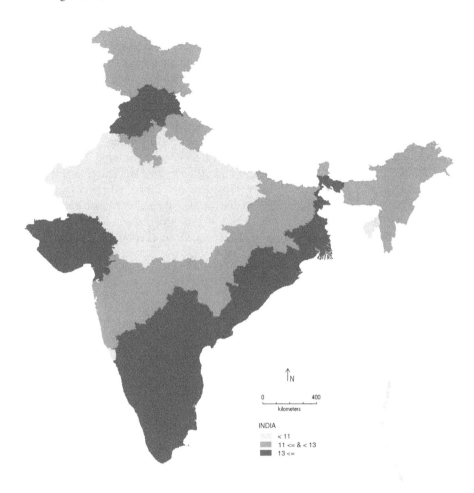

Fig. 1.7 Projected percentage of elderly population across states in India, 2026. (Source: Based on population projection report for India and states, 2001–2026. (Government of India 2006))

Workforce participation and employment patterns among the aged in India are also important to note for a variety of reasons. Employment and workforce data give information about income and income security, which shows a high level of inter-regional variation among the aged in India. Labour force participation rates among the elderly are markedly higher in rural (47%) than in urban areas (29%); were much higher for males than for females (as has been mentioned earlier); and varied significantly across states, from 26.8% in Goa to 53.5% in Nagaland. As many Indian individuals experience longer life spans and better health, an increasing number may choose to continue to work, full time or part time, beyond traditional retirement ages in order to maintain a regular flow of income without which they may not be able to keep up with their increasing expenditure on necessities such as healthcare. Analysis of the National Sample Survey data reveals that about 45% of the rural elderly are suffering from NCDs. Cough and problem of joints are the most common physical health problems among rural older people. As far as physical disabilities are concerned, in the rural areas, 5.4% of all the elderly (6.8% females and 4.4% males) are physically disabled (Alam et al. 2012).

A survey conducted in rural Rajasthan, Chhattisgarh, Gujarat, and Madhya Pradesh to study the impact of social welfare programmes on senior citizens have reported that healthcare is the main concern for the elder persons living in villages. Twenty-six percent of the elderly surveyed had not visited a village doctor at all in the past year, while only 20% of them stated that they visit a doctor at least monthly on average. Over half of the aged (55%) end up going to the local village doctor, while another 27% go to a county or township hospital. Eighty percent of the older population, who were surveyed, believed that they should get healthcare more often. Since most of them had identified themselves as poor illiterate peasant farmers, they do not have the means to seek proper medical care, the most common answer for not seeking medical assistance were lack of money (29%), lack of information about where to go (15%), and lack of motivation (11%) (Shiksha et al. 2008).

1.5 Role of Aged in Indian Society

The literature on ageing has traditionally viewed the elderly as a burden who has to be taken care of by the young, either directly, or by contributing to the state exchequer through taxes (Nyce and Schieber 2005; Palloni 2005). In recent years, however, an increasingly larger proportion of the aged are taking advantage of their prolonged life cycle to break the 'grey ceiling' and remain in the workforce. This provides an opportunity for elderly workers to become an asset to the household by supplementing household income with their earnings (WRVS 2011; Barrett and Mosca 2013; Kohli and Kunemund 2013). The financial contribution of the aged workers not only improves the economic status of the household (Selvaraj et al. 2011; UNFPA 2012), but can even reduce the poverty level.

Researchers have reported that elderly people may provide support to their families and to society in many ways (Hermalin et al. 1998), though this contribution is not always acknowledged (Help Age International 1999; WRVS 2011). WRVS (2011) reports that the elderly people help in national development by providing tax revenue to the government. The study also mentions the financial contributions made by the elderly in the form of gifts, donations and bequests to charities, child care services for families and neighbours, savings for grandchildren, and transfer of financial assets to family members. In a study of the aged in London, Barrett (2013) observes that they are substantially contributing to the society not only through their paid work but also through unpaid work—by providing care for adults, care for children, and volunteering. Support from parents to their adult children help the latter to balance between work and parenting (Kohli and Kunemund 2013). The availability of grand-parental support has been reported to increase the probability of mothers' working (Posadas and Vidal-Fernandez 2013; Husain and Dutta 2015).

According to Planning Commission (2011) the elderly people in India in many cases are the repositories of knowledge, experience, culture, and religious heritage besides bringing up grandchildren, doing voluntary service, caring for the sick and often counsel and resolve conflict by virtue of their position.

In particular, the financial contribution of elderly to their families has been found to be very important (Rendall and Speare 1995). Kohli and Kunemund (2013) have claimed that in many European countries the transfer of resources from parents to children is more frequent and more intense than the reverse flow. Kreager and Schröder-Butterfill (2003) reported that, in Java, the financial contribution of elderly people protects the family members during periods of crisis.

Most of the research on elderly in India has focussed on demographic issues, reported health status, and living arrangements, so that assessing the financial contribution of elderly has remained a largely unexplored area. Using NSSO data for 2004–2005, Selvaraj et al. (2011) estimated that elderly workers contribute 4–5% of total household expenditure. A recent study undertaken by UNFPA (2012) in seven Indian states (Himachal Pradesh, Kerala, Maharashtra, Orissa, Punjab, Tamil Nadu, and West Bengal) reports that more than half of the elderly *believe* that they make a contribution in the household budget. In fact, about one-third felt that their contribution covered more than 80% of the household budget.

Despite acknowledging the financial contribution of the aged, these studies suffer from some limitations. For instance, Selvaraj et al. (2011) failed to consider the impact of recent rural employment generating schemes like National Rural Employment Guarantee Programme providing part-time employment opportunities for the rural elderly. Nor does it examine spatial variations in the degree of contribution made by the elderly. Although this issue is partially addressed in the UNFPA study—which examines spatial variations across seven major states—the study is based on *perceptions* of contribution and does not attempt to measure the *actual* contribution made. The studies provide only (sample) estimates without statistically testing whether the level of contribution is significantly positive. Finally, the impact on poverty estimates has not been considered in either of these studies. The present study attempts to address these deficiencies.

1.6 Research Problem

In the study we will seek to examine the extent to which the aged are active in India. In particular, we will examine the extent of their labour market engagement. This will enable us to estimate the magnitude of financial contribution of elderly workers to their families in India, with special reference to Eastern India, and the role of this contribution in providing short- and long-run income stability of their households. We will also analyse time sue data of the aged to assess the extent of integration with economic and social domains and examine in what sense ageing is a gendered experience.

1.6.1 Conceptual Framework

The study is an extension of Ando-Modigliani's life-cycle hypothesis (LCH) (Ando and Modigliani 1963). This is a model based on the proposition that individuals plan their consumption and savings behaviour over their life cycle to ensure stable consumption patterns.

The LCH begins with the observation that consumption needs and income are often unequal at various points in the life cycle. Younger people tend to have consumption needs that exceed their income. Their needs tend to be mainly for housing and education, and therefore they have little savings. In middle age, earnings generally rise, enabling debts accumulated earlier in life to be paid off and savings to be accumulated. Finally, in retirement, incomes decline and individuals consume out of previously accumulated savings (Fig. 1.8).

Empirical studies of the LCH have focussed on the motivation of savings and the extent to which elderly dis-save. However, Ando-Modigliani failed to incorporate the advances in medical technology prolonging life cycles and the disintegration of traditional family-based systems providing support to the elderly (Fig. 1.9).

The prolonging of the life cycle implies, a priori, that the elderly will have to save more (and reduce consumption) during their middle age than what the LCH anticipated—with those unable to save becoming a burden to their family and society. However, the proposed study hypothesizes that prolonging of the life cycle is also an opportunity. The working period of a person can be extended beyond the conventional 'grey ceiling'. As a result, the aged population can make a financial contribution to their families. In the case of low-income households, the supplementary income of the aged can play an important role in providing short- and long-run stability to the economic fortunes of their family. In the short run, income of aged workers can help to overcome seasonal crisis

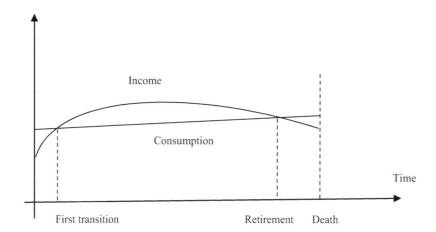

Fig. 1.8 Consumption and income trends as hypothesized in Ando-Modigliani life cycle hypothesis

1.6 Research Problem

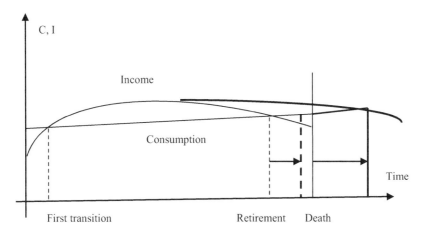

Fig. 1.9 Consumption and income trends in an ageing society

often faced by low-income families, preventing their fall into transitory poverty. In the longer run, the contribution of the aged can also push their families permanently above the poverty line.

The study argues that elderly workers can contribute by participating in economic activities. The problem with estimating their income from such activities is that elderly workers tend to be concentrated in the informal sector, working in household enterprises as unpaid family labour. Estimating the earnings of such workers will not be easy and will have to be based on imputations. Second, the aged often undertake household chores like child care, cooking, and other activities. This frees younger household members, particularly female members, who can join the work force. The imputed value of such services should also be incorporated into estimates of economic contribution of the aged.

Such estimates, however, measure *gross* contribution of the aged. It is also necessary to consider the expenditure on such persons—particularly as expenditure on healthcare of aged may be quite high. Subtracting the expenditure on aged persons will give us the *net* economic contribution of the elderly population.

1.6.2 Research Questions

The primary research question of the study is **Do the aged age actively in India?** This can be decomposed into the following questions:

1. What is the extent of labour market engagement of elderly?
2. What is the nature and gender difference in such engagement?
3. What is the net economic contribution of the elderly?
4. Does it vary across expenditure group, or household type, or socio-religious groups?

5. What are the determinants of gross and net financial contribution of elderly?
6. What is the time use pattern of aged? Are there gender differences in time use patterns? This provides information on the extent to which active ageing is occurring.

1.6.3 Database and Methodology

The first part of the study used National Sample Survey Office unit-level data from the 68th Round survey on 'Employment and unemployment'. The advantage of this data set is that it provides a nationally representative sample of households. The NSSO data, however, have three problems:

(a) It does not collect information on earnings of workers engaged in the informal sector. This is a major methodological issue as an estimated 89% of elderly workers are engaged in the informal sector.
(b) Imputed earnings of elderly persons from activities like child care is also not provided in NSSO.
(c) The expenditure data relate to the household, not to individuals.

To overcome these limitations, we undertook a questionnaire-based primary study of households with elderly persons in the capital cities of three states of Eastern India. The choice of this geographical region, comprising of states of West Bengal, Jharkhand, and Orissa, is guided by the following considerations:

- West Bengal and Orissa are states where ageing is occurring rapidly.
- The high outmigration rates of young male workers in Jharkhand weaken traditional support structures in such states.
- Orissa has a high proportion of tribal population.

We used stratified purposive sampling method (with household income and educational level being used to form the strata) to cover about 500 households from the above-mentioned states. The data were entered using a specially designed package in CSPRO. Three methods of analysis were used:

- Statistical: Estimation of gross and net contribution of aged, time use patterns, and their variations over economic-demographic factors using STATA 14.
- Spatial mapping: Using ARC-DIVA net contribution was mapped over states to examine regional patterns, if any.
- Econometric: Identify determinants of net contribution and time use pattern using appropriate econometric models using STATA 14.

1.7 Implications

Population ageing has a profound impact on societies. It affects labour markets, social security, healthcare, long-term care, and the relationship between generations. This has led to the 'silver Tsunami' being viewed as a challenge to society and the state. The concept of active ageing, as propounded in the Madrid Plan, argues that the challenge of ageing should be viewed as an opportunity. The proposed study focuses on how the aged can contribute in crucial economic ways to their households and to society as a whole. The acknowledgement of the value of the aged as an asset, rather than as an economic burden, will help to change the attitude of society and their family members towards the elderly and facilitate investment in resources and efforts to mainstream the aged in society.

1.8 Structure of Study

Chapter 2 of the study reviews the literature on contribution of the aged to society; it also contains a discussion of the findings on work force participation of the aged in India. In Chap. 3 we describe the profile of the sample used in our analysis. This is followed by findings and discussion. In Chap. 4 we use NSSO data to examine the economic contribution of the aged to their families. This is followed by an exercise examining to what extent this economic contribution has reduced the incidence and intensity of poverty in India. Regional variations are also examined using spatial techniques. From Chap. 5 onwards we use data from our primary survey undertaken in eastern India. We start by examining the economic contribution, followed by analysis of the data on non-economic contribution of aged in Eastern India. The latter is then converted to money terms to obtain estimates of total contribution of the aged to their families. We examine variations in such contribution over different socio-demographic groups using exploratory and confirmatory methods. In Chap. 6 we examine data from the time use diaries of the respondents. This is used to analyse to what extent the society are integrated with society and their families, so that they age actively. The study closes with a chapter summing up major findings and drawing policy recommendations from them.

References

Agarwal, Arunika, Alyssa Lubet, Elizabeth Mitgang, Sanjay Mohanty, and David E Bloom. 2016. *Population Aging in India: Facts, Issues, and Options*. Discussion Paper No. 10162. Bonn.

Alam, Moneer, K.S. James, G. Giridhar, K.M. Sathyanarayana, Sanjay Kumar, S. Siva Raju, T.S. Syamala, Lekha Subaiya, and Dhananjay W. Bansod. 2012. *Report on the Status of Elderly in Select States of India, 2011*. New Delhi: United Nations Population Fund.

Ando, Albert, and Franco Modigliani. 1963. The "Life Cycle" hypothesis of saving: Aggregate implications and tests. *Source: The American Economic Review* 53: 55–84.
Barker, David. 1980. Fetal origins of adult disease. *BMA Board of Science*: 149–168.
Barrett, Alan, and Irene Mosca. 2013. Social Isolation, Loneliness and Return Migration: Evidence from Older Irish Adults. *Journal of Ethnic and Migration Studies* 39. Routledge: 1659–1677. https://doi.org/10.1080/1369183X.2013.833694.
Bloom, David E., Ajay Mahal, Larry Rosenberg, and Jaypee Sevilla. 2010. Economic security arrangements in the context of population ageing in India. *International Social Security Review* 63. Blackwell Publishing Ltd: 59–89. https://doi.org/10.1111/j.1468-246X.2010.01370.x.
Bloom, D E, Cafiero-Fonseca, E M Mcgovern, K Prettner, A Stanciole, J Weiss, S Bakkila, and L Rosenberg. 2014. *The macroeconomic impact of non-communicable diseases in China and India: Estimates, projections, and comparisons*. 19335. *The Journal of the Economics of Ageing*. NBER. Belfast. https://doi.org/10.1016/j.jeoa.2014.08.003.
Centre for Ageing Research and Development in Ireland. 2011. *Global ageing: An overview*. Dublin: Centre for Ageing Research and Development in Ireland.
Desai, Sonalde B., Amaresh Dubey, Brij Lal Joshi, Mitali Sen, Abusaleh Shariff, and Reeve Vanneman. 2010. *Human development in India: Challenges for a society in transition*. New Delhi: Oxford University Press.
Dey, Subhojit, Nambiar Devaki, J.K. Lakshmi, Kabir Sheikh, and K. Srinath Reddy. 2012. Health of the elderly in India: Challenges of access and affordability. In *Aging in Asia: Findings from new and emerging data initiatives*, 371–386. Washington, DC.
Flatt, Thomas. 2012. A new definition of aging? *Frontiers in Genetics* 3: 1–2. https://doi.org/10.3389/fgene.2012.00148.
Giridhar, Garimella, K.M. Sathyanarayana, Sanjay Kumar, K.S. James, and Moneer Alam. 2014. Population ageing in India. In *International Journal of Ageing and Later Life*, vol. 10. New Delhi: Cambridge University Press: 73–77.
Government of India. 2006. *Population projections for India and states 2001–2026: Report of the technical group on population projections*, 287. New Delhi: Office of Registrar General and Census Commissioner, Government of India.
———. 2011. Census of India 2011.
Help Age International. 1999. *Older people in disasters and humanitarian crises: Guidelines for best practice*. London: Help Age International.
Hermalin, Albert, Yi-Li Chuang, Zachary Zimmer, and Xian Liu. 1998. Educational Attainment and transitions in functional status among older Taiwanese. *Demography* 35. Springer: 361. https://doi.org/10.2307/3004043.
Hornby, Albert Sydney, Joanna Turnbull, Diana Lea, Dilys Parkinson, Patrick Phillips, and Michael Ashby. 2010. *Oxford advanced learner's dictionary of current English*. 8th ed. Oxford: Oxford University Press.
Husain, Z., and M. Dutta. 2015. Grandparental childcare and labour market participation of mothers in India. *Economic and Political Weekly* 50: 74–82.
Institute for Health Metrics and Evaluation. 2013. *The global burden of disease*. Seattle: Institute for Health Metrics and Evaluation.
Kohli, Martin, and Harald Kunemund. 2013. The Social connections of older Europeans. In *The SAGE handbook of aging, work and society*, ed. John Field, Ronald J. Burke, and Cary L. Cooper, 347–362. London: SAGE.
Kreager, Philip, and Elisabeth Schröder-Butterfill. 2003. *Actual and de facto childlessness in East Java: A preliminary analysis*. Oxford Institute of Ageing Working Paper no WP203. Oxford: Oxford Institute of Ageing.
Kübler-Ross, Elisabeth. 1969. *On death and dying : What the dying have to teach Doctors, Nurses, Clergy and their own Families*. New York: Macmillan.
Nyce, Steven A., and Sylvester J. Schieber. 2005. *The economic implications of aging societies: The costs of living happily ever after*. Cambridge: Cambridge University Press.

Palloni, A. 2005. The influence of early conditions on health status among elderly Puerto Ricans. – PubMed – NCBI. *Soc Biol.* Fall-Winte: 132–163.
Posadas, Josefina, and Marian Vidal-Fernandez. 2013. Grandparents' Childcare and Female Labor Force Participation. *IZA Journal of Labor Policy* 2. SpringerOpen: 14. https://doi.org/10.1186/2193-9004-2-14.
Rendall, Michael S., and Alden Speare. 1995. Elderly poverty alleviation through living with family. *Journal of Population Economics* 8. Springer: 383–405. https://doi.org/10.1007/BF00180875.
Riley, Matilda White. 1978. Aging, Social Change, and the Power of Ideas. *Daedalus. The MIT PressAmerican Academy of Arts & Sciences.* https://doi.org/10.2307/20024579.
Rose, Michael R., Thomas Flatt, Joseph L. Graves, Lee F. Greer, Daniel E. Martinez, Margarida Matos, Laurence D. Mueller, Robert J. Shmookler Reis, and Parvin Shahrestani. 2012. What is Aging? *Frontiers in genetics* 3. Frontiers Media SA: 134. https://doi.org/10.3389/fgene.2012.00134.
Selvaraj, S., Y. Balarajan, and S.V. Subramanian. 2011. Health care and equity in India. *The Lancet* 377. Elsevier: 505–515. https://doi.org/10.1016/S0140-6736(10)61894-6.
Shiksha, Sonali Public, Punjabi Mohalla, Gurudwara Road, and Guna Madhya Pradesh. 2008. *Study report on a study of effectiveness of social welfare programmes on senior citizen in rural Madhya Pradesh Planning Commission Government of India.* Gunai.
Subrahmanya, R. K. A. 2017. *Social security for the elderly in India.* Thematic Paper 1. New Delhi.
Suzman, Richard, and John Beard. 2011. *Global Health and aging.* Vol. 1. Baltimore: NIH Publication no 117737.
United Nations, Department of Economic and Social Affairs, Population Division. 2015. *World Population Ageing* 2015. United Nations, New York (ST/ESA/SER.A/390).
———. 2017. *World population ageing.* New York: United Nations.
UNFPA. 2012. *By choice, not by chance: Family planning, human rights and development.* New York: UNFPA.
———. 2017. *Caring for our Elders: Early responses caring for our elders: Early responses.* New Delhi: UNFPA.
Uppal, Sharanjit, and Sisira Sarma. 2007. Aging, health and labour market activity: The case of India. *World health & population* 9: 79–97.
Whitbourne, Susan Krauss, and Stacey B. Whitbourne. 2010. *Adult development and Aging: Biopsychosocial Perspectives.* New Jersey: Wiley.
WHO. 2015. *WHO | What is healthy ageing? WHO.* Geneva: World Health Organization.
World Health Organization. 2002. *Active ageing: A policy framework.* Geneva: World Health Organization.
WRVS. 2011. *Valuing the socio-economic contribution of older people in the UK.* Cardiff: WRVS.

Chapter 2
Positive Implications of Ageing for Indian Families: Aged as an Asset

2.1 Introduction

As countries age, assigning a societal role to the aged emerges as a major challenge. Research on ageing in developing countries focuses on the role of family in supporting and meeting the needs of the aged and mourns the disintegration of the traditional joint family and the decay of traditional norms relating to the care of the aged. This implicitly assumes that the aged is a burden on the family, who have to be looked after and supported, but does not actively take part in society or family. In fact, the noted demographer, Paul Demeny—while delivering the plenary lecture at the European Population Conference in Budapest, 2016—even referred to ageing as a tax on the younger members of society. In recent years, however, literature on gerontology argues in favour of a more active role for the aged. In particular, the concept of active ageing has emerged as an important proposition.

2.2 Active Ageing

The concept of active ageing was introduced by WHO (2002). The report commences by arguing that if ageing is to be a positive experience, the increase in longevity must be simultaneously accompanied by an increase in opportunities for health, participation and security. Active ageing is the process to achieve this objective. It is defined as '… the process of optimizing opportunities for health, participation and security in order to enhance quality of life as people age' (World Health Organization 2002).

The key aspects of active ageing are as follows:

1. ***Autonomy***: Perceived ability to control, cope with, and make personal decisions about how one lives on a day-to-day basis, according to one's own rules and preferences.

2. ***Independence***: This relates to the ability to perform functions related to daily living with no and/or little help from others.
3. ***Quality of life:*** This is based on the individual's perception of his or her position in life in the context of the culture and value systems where they are situated. It is a broad concept, incorporating a person's physical health, psychological state, level of independence, social relationships, personal beliefs, and relationship to salient features in the environment. As people age, their quality of life is largely determined by their ability to maintain autonomy and independence.
4. ***Healthy life expectancy:*** This relates to how long people can expect to live without disabilities.

In order to achieve active ageing, certain conditions must be satisfied. One of the most important of the conditions is:

> When labour market, employment, education, health and social policies and programmes support their full participation in socioeconomic, cultural and spiritual activities, according to their basic human rights, capacities, needs and preferences, people will continue to make a *productive contribution to society in both paid and unpaid activities* as they age. (World Health Organization 2002: 46, emphasis added)

The present study takes the italicized phrase as the starting point of our analysis. The report seeks to examine the extent of participation in the labour market so that the aged can contribute to the family in terms of money. It also recognizes that the aged may also provide unpaid family labour within the household. This can also improve family outcomes. While studies in India have focused on the former, the current study also examines the non-economic dimension of the contribution of the aged. This is an important contribution of the present study.

2.3 Work Participation of Elderly: A Brief Literature Review

A review of the literature on aging in India reveals a substantial number of studies on demographic issues, health status, social security, and ill treatment (Husain and Ghosh 2010; Rajan and Mishra 2011; UN 2013). There are comparatively few studies on participation of the Indian elderly in the labour market (exceptions are Pandey 2009; Rajan et al. 2003; Selvaraj et al. 2011; Alam and Mitra 2012; Singh and Das 2012; UN 2013; Dhar 2015).

2.3.1 Workforce Participation Among the Aged in India

Analysing Census data, Rajan et al. (2003) found that the workforce participation (WFP) of the elderly in India has decreased from 1961 to 1991; this was observed for both rural and urban parts of India. In comparison with the elderly in urban

India, the elderly in rural India participated more in the workforce. Disaggregating the analysis by gender, the study found that the elderly male participated more in economic activities than the female elderly. The study also found that the share of the elderly workers engaged in agriculture has increased over the study period, with nearly 80% of the elderly workers being engaged in this sector in 1991. Among the male elderly, 62% worked as cultivators, whereas among the female elderly, 70% worked as agricultural labourers.

Using National Sample Survey (NSS) data from 1983 to 2004–2005, Selvaraj et al. (2011) analysed the elderly workforce participation trend in India on the basis of usual activity status (usual principal status[1] and usual subsidiary status[2]). They have shown that the elderly workforce participation rate (WFPR) has decreased marginally in the post-globalized period compared to pre-globalization (from 42% in 1983 to 39% in 2004–2005), mainly due to growing number of elderly in the above 80 years and above (the so-called elderly old) age group who are less likely to participate in the workforce. The total number of elderly workers in India was estimated to be 31 million in 2004–2005; this is approximately 7% of the total workforce (Selvaraj et al. 2011). Although the WFPR of the elderly is higher in rural areas compared to urban areas, elderly employment shows a fluctuating trend in rural India. In urban India, on the other hand, WFPR declined significantly from 31% to 23% over the study period. The work participation of the male elderly declined from 64% in 1983 to 57% in 2004–2005, while female employment trend remains stagnant at 20% in both the years. Disaggregating the analysis by place of residence and gender, Selvaraj et al. (2011) found that for urban male, WFPR declined sharply (from 50.2% in 1983 to 36.6% in 2004–2005), while for rural male it declined slightly (66.8–64.4%). WFP for the rural females increased marginally from 22.6% to 25.3%, while WFP of urban females declined from 13.8% to 10%. The analysis also reveals that the workforce participation decreases with age—with the rate of decrease being higher for female elderly.

Selvaraj et al. (2011) also analysed the educational profile of aged workers. More than 70% of elderly workers are illiterate or do not have any primary education. Among the female workers, illiteracy is even higher—almost 93%. The proportion of illiterate elderly workers is higher in rural areas as compared to the urban areas. Selvaraj et al. (2011) also found that the labour force participation of the elderly is higher among the poor consumption quintiles than the richer ones, particularly among the female elderly workers. These findings indicate that it is economic vulnerabilities that is motivating the aged to work in India.

Most of the elderly workers are self-employed—almost 78% of all elderly workers. Casual employment is comparatively higher among the female elderly. In urban areas, a significant proportion of elderly female workers is engaged in regular

---

[1] If an individual is identified as a worker for the major part of the year, he/she is categorized as a worker on the basis of the usual principal status.

[2] If an individual is identified as a worker only for a minor part of the year, he/she is categorized as a worker on the basis of subsidiary status.

employment. Analysis of current weekly status[3] data of NSS has also shown that the real wage of regular and casual workers has increased by 60% between 1983 and 2005. Selvaraj et al. (2011) also observed that although the elderly are receiving wages lower than that of non-elderly workers, the contribution of the former to the total household income is important—amounting to about 4–5% on an average.

UNFPA (2012) have carried a survey regarding the socio-economic status, health status, and work participation, social security schemes of the elderly people in seven states (Himachal Pradesh, Kerala, Maharashtra, Orissa, Punjab, Tamil Nadu and West Bengal) in India in 2011. Among the states, Maharashtra has the highest elderly work participation rate followed by Orissa, while Kerala has the lowest elderly work participation rate. They have observed that the work participation of the male elderly is significantly higher (39%) than the female (11%) and rural elderly people are participating more in the workforce. Most of the workers are in the 60–69 age group. Among the elderly, the participation of the oldest old men (80+) is also quite high (13%). UNFPA (2012) have found that the elderly who are currently working, work full time. The study also reported that the poor and the less educated people are participating more out of their economic compulsion. The elderly who are never married, divorced, or separated have higher work participation. The high workforce participation rate is also observed among the Schedule Castes (SCs) and Schedule Tribes (STs). Regarding the sector of employment, it has been noticed that most of the elderly are either self-employed or working in the informal sector, and only 5% of the elderly work in some temporary position in the public sector (UNFPA 2012). Although the income earned by the elderly is not high, according to the UNFPA study, they are making a regular contribution to the daily household expenditure.

The relation between work participation and health of the elderly people has also been examined in the Indian context. Alam and Mitra (2012) focused on the adverse effects that work has on health, while Pandey (2009) found that bad health lowers workforce participation.

2.3.2 Work and Earnings

Singh and Das (2012) have used information provided by NSS on current weekly status data from 1993–1994 to 2009–2010. They have defined the work participation of the elderly on the basis of wage received by them—an elder is considered part of the labour force only if he/she gets wage on any of the 7 days in the week preceding the date of survey. Singh and Das (2012) have considered weekly days of work of an elderly as a measure for weekly labour supply. Accordingly, if an elder works for a full day, it was taken as 1 day, and if an elder works for half day, it was

---

[3] It is the activity status obtained for a person during a reference period of 7 days preceding the date of survey.

2.3 Work Participation of Elderly: A Brief Literature Review

taken as 0.5 days. Singh and Das (2012) found that, while wage labour participation of the elderly has decreased in urban areas (from 7.45% to 6.01%), it has increased in rural areas (9.66–11.35%) from 1993–1994 to 2009–2010. The average weekly days of work supplied by the working elders has decreased in rural areas (from 6.22% to 5.80%) but has remained same in urban areas (6.42%). The study also reported that elders from small families are participating more in the workforce—an important trend, given the decline in family size in India. The wage labour participation of elderly Hindus has increased over time, while that of elderly Muslims has decreased in rural India. In rural India, the proportion of agricultural and other elderly labour has increased (from 83.6% to 86.8%) and, in urban areas, the proportion of casual labour has increased (from 37.88% to 41.07%).

Singh and Das (2012) also undertook an econometric analysis of determinants of the decision to work. The impact of variables like age, square of age, marital status, sex, education, caste and religion, number of children (below 18 years), number of adults (18 years and above) in the household, amount of land cultivated in acres, and monthly per capita expenditure on workforce was considered. The analysis was based on a probit model. Singh and Das (2012) found that in urban areas, there is a negative relationship between the probability of wage labour participation and age of the elderly. A similar result is obtained in rural areas for the period 1993–1994; for 2009–2010, the coefficient of age was found to be insignificant. In rural areas, the aged from Schedule Caste (SCs) and Schedule Tribes (STs) and in urban areas the aged from SCs are more likely to be working than other castes in 2009–2010. Compared to the elderly males, female elderly workers participate less in rural and urban areas in both the rounds. Singh and Das (2012) found that the elderly from poorer households have a higher probability of wage labour participation in both the rounds. This holds true for both rural and urban areas. Singh and Das (2012) report that educational status does not affect wage labour participation; only in urban areas is workforce participation rates of secondary and higher educated persons significantly higher than that of illiterates. Another important finding is that elders from smaller families are more likely to participate in work than elders from larger families.

Using Heckman sample selection regression, Singh and Das (2012) found that in 2009–2010, in rural and urban areas, the weekly days of work supply by the working population of the elderly do not have significant relationships with their age. Singh and Das (2012) argued that in urban area, the elderly SCs show a lower weekly number of days of labour supply in 2009–2010 compared to others. In contrast, in both rural and urban areas, Muslim elderly are working more days than the Hindu elderly. In urban areas, female elders work a lesser number of days as compared to male elders. Education does not have any major significant effect on weekly days of work supplied in rural as well as in urban areas. Elderly from smaller households are found to supply a significantly lesser number of days of weekly work compared to larger families in both the rounds.

2.4 Aged in the Labour Market

Four recent contributions to the literature on workforce participation of the aged and its patterns in India are by Dhar (2014, 2015, 2017) and Dhar et al. (2015). In this section, we summarize the findings of these works.

2.4.1 Workforce Participation

WFPR of the elderly has declined from 35.40% in 1999–2000 to 34.40% in 2011–2012 (a marginal decline of one percentage point). The declining participation of the aged in the labour market is likely to be the result of fewer job opportunities, lack of skills commensurate with modern production techniques, and poor health status. Another possible reason is that aged persons give up hope of finding jobs once they become unemployed—they withdraw from the labour market on losing their job.

A regression model was also estimated to verify whether the aged are participating less, even after controlling for socio-economic variables, in the labour market. Since the model was characterized by endogeneity, Arellano's control function approach was used to estimate it (Arellano 2008). Results indicate that WFPR has declined significantly in the rural sector. It is possible that government policies like loan waiver scheme, Mid-Day Meal scheme, MNREGA and public distribution system have protected the Indian economy from the adverse effect of the global economic crisis and the 2009 drought. So, the results may reflect the success of the above-mentioned government programmes. However, Chandrasekhar and Ghosh (2007) have argued that greater use of mechanized techniques in the field of agriculture has reduced the demand for labour per unit of output. Given that unemployment rate of the elderly population is marginal in both rounds, the declining WFPR is also likely to be a result of a withdrawal of the aged from the workforce. In contrast, there has been no significant change in WFPR among elderly urban males over the period of study.

Analysis of WFPR by education and per capita household expenditure reveals that it is the less educated and poorer respondents who are participating more in the labour market. This implies that it is poverty that drives the aged to work. Econometric analysis also reveals that elderly persons from socially marginalized communities like Scheduled Castes (SCs), Scheduled Tribes (STs), and Other Backward Castes (OBCs) are, in general, more likely to work, compared to Hindu Forward Castes (HFCs). In the case of Muslims, results differ between rural and urban areas. While, in rural areas, Muslims are less likely to find work as compared to HFCs, the opposite is observed in urban areas.

2.4.2 Informal Sector and the Aged

In India, the informal sector is the largest employment providing sector. Given that the population and workforce now contain a greater share of aged persons, we would expect that the extent of informalisation will increase over time. As expected, we find that there is a positive relationship between age and informal sector participation in both the rounds (Fig. 2.1). This is not surprising, what is important is that informalization has increased over the study period for most of the age groups.

The increasing informalization of the elderly workforce may simply be due to the fact that full-time employment in the public sector is possible only up to 60, or at the most 65 years. This implies that employment opportunity in the formal sector is limited for such workers, and also increasing the proportion of 'middle-old' (persons aged 70–79 years) in the population and workforce should result in an increased proportion of informal sector workers among the elderly. Our analysis shows that the informal sector participation of the elderly workers aged 60–64 years increases by 2% (from 74% to 76%), among the 65–69 years aged from 78% to 79%, and among the 70–74 years aged from 82% to 83%. The highest increase has been noticed in the 75–79-year age group (increases from 88% to 93%). The proportion of workers aged 15–59 years engaged in the informal sector has grown and such increase varies from 5% to 7% over the study period.

While this result may be attributed to jobless growth in the Indian economy, squeezing out the workers from the formal sector, such an explanation overlooks the trends in employment in India during the study period. Studies have shown that while the growth rate of the organized sector employment declined from 0.4% per annum during 1994–2000 to −1.1% in 2004–2005, it subsequently increased to 0.7% in 2005–2008. Several studies have also reported organized manufacturing

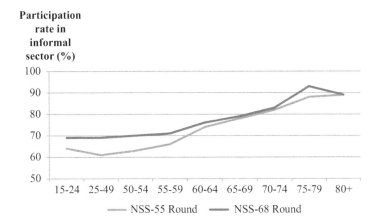

Fig. 2.1 Informal sector participation of different age groups in India in the 55th and the 68th rounds of NSS. (Source: Estimated from NSS data)

sector to have increased at the rate of 7.5% per annum between 2003–2004 and 2008–2009.

If we consider the traditional view of informal sector, then increasing informalization may be a sign of concern. This is because increasing concentration in this sector implies quality of employment has deteriorated over the study period; further, studies have shown that there is a link between informal sector participation and poverty. But the informal sector has also been shown to evolve over time, so that it now has greater linkage with the formal sector. What is more likely, therefore, is that the increasing integration of the formal and informal sectors has led to the creation of job opportunities and an increase in real earnings in the latter. Mentioned that the relation between formal and informal sector is not anti-cyclical but pro-cyclical; whenever formal economy expands, subcontracting will expand the informal sector. Given the easy nature of entry into the informal sector labour force, this has led to aged workers from low-income households flowing to this sector to augment household income.

Our analysis also reveals that it is the more educated aged persons, from affluent households, who are likely to participate in the informal sector. On the other hand, there has been a decline in the share of urban women workers from the bottom expenditure quintile group in the informal sector. This may reflect their inability to continue working because of their poor health status. Another possibility is that such women may shift to household activities like looking after their grandchildren, cooking, and similar chores, thereby facilitating the entry of younger and more productive women into the labour market.

We have also undertaken an econometric analysis, using a probit model, to identify determinants of being in the informal sector vis-à-vis the formal sector. A time dummy is incorporated to examine whether the probability of being in the informal sector has gone up or down over the study period. The coefficients of the Time dummy are significant and negative for all four sub-samples—rural males, rural females, urban males, and urban females. The decline is greatest for rural females. This is possibly because they are unable to compete with other workers—both younger females and elderly males—who are more capable of adjusting themselves to the demands of the technology and organizational forms emerging in the informal sector.

Another important finding is that the workers from disadvantaged castes are participating less in the informal sector than the Hindu upper class. This may be because of exclusion from the informal sector. However, the probability of a Muslim male participating in the informal sector is significantly higher than that of a Hindu upper caste worker in urban areas. This is in keeping with studies of the Muslim community. Muslim female elderly are also participating more in this sector compared to Hindu upper caste and like male elderly over time participation increases.

2.4 Aged in the Labour Market

Table 2.1 Occupational segregation between the elderly and near elderly in the 55th and the 68th rounds of NSS

Group	Duncan Index		Hutchens Index	
	NSS 55th	NSS 68th	NSS 55th	NSS 68th
Rural male	0.17	0.15	0.03	0.02
Rural female	0.06	0.08	0.01	0.01
Urban male	0.27	0.23	0.06	0.04
Urban female	0.24	0.24	0.09	0.07

Source: Calculated from NSS 55th round and NSS 68th round

2.4.3 Occupational Pattern of Aged Workers

Our study has also examined the quality of employment; this is reflected in the occupational pattern of the elderly. Two questions are important in this context: is there any occupational segregation in the occupational choice between the elderly and near-elderly (50–59 years) workers? The absence of segregation would indicate that workers are probably remaining in the same occupation after crossing 60 years. In the second step, we will identify the sectors where aged workers are concentrated and examine whether these are high earnings or low earnings sector.

To analyse the pattern of occupation, we have used the National Classification of Occupation (NCO) codes. However, there is a problem in using this code over rounds. While the 55th round of NSS has followed NCO 1968 classification of occupation, the NSS 68th round has followed the NCO 2004 classification. To ensure consistency between rounds, concurrence table was used to convert the 1968 codes to the NCO 2004 classification. We have clubbed some occupations and deleted some others to arrive at a uniform occupational structure. We present results for the two-digit classification systems.

Table 2.1 reports occupational segregation between the above two groups of workers, elderly and near elderly, in each round—disaggregating the sample by place of residence and gender. The low value of the segregation index suggests that aged workers continue to use their skills and experience by remaining in the same occupation after 'retirement'. They only shift from the formal to the informal sector (the econometric result shows that with the increase in age, participation in the informal sector increases among the rural and urban male elderly). Given that the latter is typically an unregulated sector, this would allow employers to exploit the skill and experience of elderly workers by offering them wages below the market rate. Changes in the segregation index over time are marginal, indicating that this trend is occurring in both rounds.

We next examined whether the aged are concentrated in high-paying or low-paying sectors. This analysis was undertaken in two steps. In the first step, we analysed the occupational pattern of the elderly to identify sectors where they concentrate. We then estimated average earnings of *all* workers in each occupational category to find out whether the occupations where elderly workers concentrate are high-earning occupations or not.

We have found that 84% of rural male aged workers are participating in the primary sector (such as market-oriented skilled agriculture, and fishery work, agriculture, fishery-related works etc.). In the 68th round, the high concentration of the rural male aged in the primary sector persist but declined to 78% (a decline of six percentage point). Over the study period, the concentration of the rural male elderly workers has increased in mining, construction, and manufacturing sector. The proportion of male elderly workers working as corporate managers, extraction workers, and building trade workers in rural India have increased in 2011–2012 as compared to 1999–2000.

Analysis of earnings in different occupation shows that the earnings of the aged in the primary sector are low in both the rounds. The concentration of the rural male elderly is very low in the high-earning sector. Comparison of average earnings of the elderly and the average earnings of all workers in that particular sector illustrates that in most of the occupations, the elderly earn less than the average earnings in that occupation. But it is not necessary that in all the occupations, earnings of the elderly are less than the average earnings in that sector. In practice, we observe that in 1999–2000, average daily earnings of the elderly in some occupation (teaching profession, mining, and construction-related works and in physical, mathematical, and engineering work, etc.) is greater than the average earnings of all workers in those occupations. In 2011–2012, the same has been observed in case of physical, mathematical, and engineering work and science associates, handicraft, printing-related work, etc.

We also found that almost 86percent of rural female elderly participate in the primary sector (which is a low-earning sector) in the 55th round of NSS. In the 68th round, similar to rural males, we find that the concentration of rural female elderly in the primary sector decreases by seven percentage point. In the 55th round, rural female elderly in all occupation earn less than the average earnings in each occupation. Rural female elderly earn much less in case of life science, health profession, clerical work related to customer service, mathematics, engineering work, legislative work, etc. compared to the average earnings of all workers in those occupations. However, the concentration of aged rural female in those occupations is very negligible. In the 68th round, 43.9% of the female elderly are engaged in the market-oriented skilled agriculture and related work, and a positive sign is that in this occupation, the elderly female earn more than the average daily earnings of all workers in that occupation. Rural female elderly also earn slightly higher than the average earnings of all workers in case of agriculture and fishery-related work, where almost 30.12% aged female is involved. Although female elderly earn more when they work as machine operators and assemblers, in such jobs concentration of them is not substantial.

Most of the urban male elderly are concentrated in service and manufacturing sector in both the rounds. In 1999–2000, nearly 13% the aged males worked as corporate managers, which is a high-earning sector. A welcome sign is that in 2011–2012, the proportion of the aged in such occupation has increased to 25%. However, most of the elderly male in urban India concentrated in low-paying jobs. In the 55th round, only 16% and in the 68th round 27% elderly males on an average earn more than ₹ 200 per day. Analysis also reveals that 30% of the male aged workers in some occupations (e.g. in the model, sales and demonstrator work, craft and related works, and in teaching profession) earn more than the average earnings of all workers in those jobs in the 55th round of NSS. In the 68th round, 52% of the male elderly earn more than the average earnings of all workers in the occupations like corporate managers, market-oriented skilled agriculture and related works, extraction and building works, life science and health profession, etc.

Analysis of occupation choice of urban elderly female shows that more than 70% of aged workers are engaged in the service and manufacturing sector. This is observed in both the rounds. A very high concentration has been observed in the low-paying jobs. In the 55th round, only 0.8% female elderly in urban India are involved in the high-paying sector, and in the 68th round, we observe a little increase in the proportion (3.7%) of people who are working in the high-paying sector. Most of the urban female elderly are engaged in low-paying jobs; however, in the 55th round in personal and protective service work and clerical jobs and in the 68th round, we observe that earnings of the elderly female engaged in agriculture, fishery-related work, market-oriented skilled agricultural work and in teaching profession are higher than the average earnings of all workers in such occupations.

2.4.4 Earnings

Analysis of daily earnings by age groups broadly reveals an inverse 'U'-shaped relationship between the two in both the rounds, with earnings being highest among the 50–54 years and 55–59 years age group in the 55th and 68th rounds, respectively. In the early years of working life, people usually have lower earnings because of lack of experience, but due to the advancement of age, their skill, experience, and knowledge also increase. So, they have higher earnings in their later life. The daily average real earnings of the elderly are substantially lower than that of other age groups. The wage difference may be explained in terms of lower reservation wage of the elderly. There are several reasons for this, such as fewer work opportunities, need to work at a nearby places, and need for flexible employment hours.

Figure 2.2 also reveals that the daily average real earnings of young workers aged 25–49 years have decreased—with the extent of change being statistically significant (at 1% level). A possible cause is the recessionary conditions in the Indian manufacturing sector (caused by global recession), accompanied by drought affecting prosperity levels in rural India. Real earnings of near-elderly workers (aged 55–59 years) have significantly increased marginally over the study period. Among the elderly workers, those aged 75–79 years have experienced a significant reduction in daily average real earnings. Elderly workers who are less competitive, have lower reservation wages, and have experienced a significant reduction in earnings are expectedly worse hit by the resulting contraction in the labour market due to recessionary conditions.

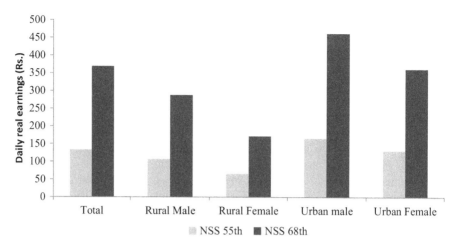

Fig. 2.2 Daily average real earnings of the elderly in the 55th and 68th rounds of NSS. (Source: NSS 55th and 68th round)

We next concentrate on the gender-wise changes in real earnings of elderly workers. Figure 2.2 shows the daily average real earnings of all elderly, male elderly, and female elderly workers in rural and urban India in the 55th and 68th rounds of NSS. Figure 2.2 illustrates that the mean daily real earnings of the elderly workers have increased over the study period; further, this increase has occurred for both male and female elderly workers in rural India, though the increase is greater for rural female workers. In urban India also female elderly workers have experienced an increase in earnings, but the real earning of male elderly workers has decreased. Expectedly, daily earnings of female elderly in both rural and urban India are much lower than the male elderly in both the rounds. The results match with the findings of the existing studies showing that, among the regular and casual elderly workers, female workers are getting lower wages than the male workers in India. In particular, two reasons may be identified for this phenomenon:

(a) Lower earnings of elderly women may be due to discrimination and occupational choice by women into lower paying jobs.
(b) Another reason may be the traditional perception that older women are physically weaker than men.

A Heckman selection model was also estimated to determine the change in daily earnings over time. The coefficient of the Time dummy was significant at 1% level for urban male and female and rural male and female elderly workers. We find that over the study period, daily earnings of urban elderly have decreased. Rural male and female elderly have experienced an increase in earnings in recent years. This is similar to what we observed in descriptive analysis.

Econometric results also indicate that daily earnings of the Scheduled Castes in both the rounds are lower than the Hindu upper caste for all groups of the elderly except the rural male in the 68th round. Urban Muslim male elderly workers in both the rounds and urban female, rural male, and female elderly in the 55th round of NSS earn significantly lower than the Hindu upper caste. However, in the 68th round, daily earnings of urban female, rural male, and female elderly workers are higher than the Hindu upper caste. The earnings of the urban elderly Scheduled Tribe males and females and rural Scheduled Tribe females are higher than earnings of Hindu upper caste in the 68th round.

After analysing the determinants of earnings of elderly workers, we have estimated another regression considering the elderly and non-elderly workers The results reveal that the just adult (15–24 years) urban male, middle-aged (25–49 years) workers, and elderly workers are getting significantly lower earnings compared to the near-elderly (50–59 years) workers in the 55th and also in the 68th round of NSS. Analysis of earnings of urban female workers for all age groups also confirms the same results in both the rounds (except the elderly who are above 75 years). Moreover, the gap between the earnings of the elderly and near elderly has increased over the study period among the urban males as indicated by the absolute increase in the value of regression coefficients of different age dummies of urban male elderly and 60–65 years urban female elderly. However, the gap between earnings of elderly and near elderly has decreased among the rural males and females.

A decomposition analysis, following, was carried out to identify the reasons for the disparity in daily earnings. We found that the unexplained (or residual) wage gap is more than 50% in the 55th round and more than 80% in the 68th round. Discrimination between the earnings of the elderly and near elderly is highest among the rural female workers, followed by urban male workers in the 55th round. In the 68th round, discrimination is highest between elderly and near-elderly urban male workers, followed by urban female workers. A welcome development is observed over the study period—over the study period, discrimination among the rural male, rural female, and urban male elderly workers has decreased. However, among the urban females, the extent of discrimination against elderly workers is not only substantial but has also increased over the study period.

2.4.5 Non-financial Contribution of the Elderly

Apart from their financial contribution, in the form of earnings from the labour market, to their household, the aged may also provide assistance to their family members in other non-financial ways. The availability of grand-parental support has been reported to increase the probability of mothers' working. Sometimes older people maintain full parenting responsibility. This care providing rate is, however, higher among the women elderly than men and such rate declines with the increase in age of the children, for instance, in London, the grand-parental care provision has found to be highest for the 0–2 years aged children. A primary survey of married women with children below 8 years found that 49% of women with grand-parental support worked; in contrast, only 35% of women without such support worked. Further, the proportion of women working despite having elderly support is low among respondents married into conservative families (who object to women working), residents of the less liberal north Kolkata, non-Hindus (of which Muslims form the majority), backward castes (who are also likely to be conservative) and non-Bengalis (comprising mainly of north Indians and Muslims). This indicates the influence of norms and culture mediating the relationship between elderly support and the labour market impact of mothers.

In European societies "Grandparent care can take many forms, from occasional babysitting through regular help with child care to being the sole or main provider of childcare while parents work, or living with their grandchildren in multi-generation households". The dependence on grandparents is much higher in Kolkata—the majority of grandparents cook/supervise cooking, serve food/feed grandchildren, and read to/play with grandchildren.

Elderly people are found to assist not only in childcare but also in housework, which enables the younger people to participate in productive tasks mentioned that during the early 1990s in Russia and Romania, the aged people spent 20 h in a week queuing in shops, thereby freeing the younger people to do their work.

The elderly are not only committed to their family but also to the community. They ensure community safety, bonding within society and provide valuable advice, etc. Due to the advancement of level of literacy and medical science, and spread of general awareness in the near future, the older people will be healthier, more educated, and more active, which will ensure greater contribution from the elderly to the society, e.g. it has been claimed that social care provided by the elderly in the UK is currently £34 billion, and this will grow to £52 billion by 2030, and the older volunteers are providing considerable hidden value to the national economy in the UK, and this is approximately £10 billion. According to Planning Commission (2011), the elderly people in India in many cases are the repositories of knowledge, experience, culture, and religious heritage besides bringing up grandchildren, doing voluntary service, caring for the sick, and often counsel and resolve conflicts by virtue of their position. The elderly are contributing to the society through their paid and unpaid work.

2.5 Conclusion

To sum up, the rapid process of ageing in India will pose a major challenge before the Indian society and families in the coming years. It has been the practice of researchers to focus on the negative aspects of the challenge—increasing expenditure on healthcare of the aged,[4] ensuring care services for the elderly, and the issues centering on a rising elderly dependency ratio. But ageing also represents an oppor-

---

[4]Alemayehu and Warner estimate that lifetime health costs are USD 316,600, of which about half are incurred in the senior years. https://www.ncbi.nlm.nih.gov/pmc/articles/PMC1361028/

tunity. There is a large body of research that points out to how the elderly can contribute towards the welfare of their families. If the Indian society and the State can ensure that the aged are active, then it is possible for the aged to contribute more fully to their families. In such cases, they will transform from being a burden to an asset.

References

Alam, M., and A. Mitra. 2012. Labour market vulnerabilities and health outcomes: Older workers in India. *Journal of Population Ageing* 5 (4): 241–256.

Arellano, M. 2008. *Binary models with endogenous explanatory variables*. Accessed at https://www.cemfi.es/~arellano/binary-endogeneity.pdf on 23 October 2019.

Center for Health Development, World Health Organisation. 2002. *Active ageing: A policy framework*. Geneva: World Health Organisation.

Chandrasekhar, C.P., and J. Ghosh. 2007. Recent employment trends in India and China: An unfortunate convergence? *The Indian Journal of Labour Economics* 50 (3): 383–406.

Dhar, A. 2014. Workforce participation among the elderly in India: Struggling for economic security. *The Indian Journal of Labour Economics* 57 (3): 221–245.

———. 2015. *Recent changes in job opportunities for aged in India: Evidence from NSSO data*. New Delhi: Institute for Human Development: South Asian Research Network Working paper SAR/008/W/I.

———. 2017. *Work force participation of aged in India: Important determinants*. Unpublished dissertation, submitted to Calcutta University.

Dhar, A., Z. Husain, and M. Dutta. 2015. Is the bell tolling for aged workers? *Journal of Regional Development and Planning* 4 (1): 1–19.

Husain, Z., and S. Ghosh. 2010. Economic independence, family support and perceived health status of elderly: Recent evidence from India. *Asia-Pacific Population Journal* 25 (1): 47–77.

Pandey, M. K. 2009. *Labour force participation among Indian elderly: Does health matter?*. Canberra: Australia National University, Australia South Asia Research Center, Working Paper 2009/11. https://taxpolicy.crawford.anu.edu.au/acde/asarc/pdf/papers/2009/WP2009_11.pdf. Accessed 10 Feb 2019.

Rajan, S.I., P.S. Sarma, and U.S. Mishra. 2003. Demography of Indian aging, 2001-2051. *Journal of Aging and Social Policy* 15 (2–3): 11–30.

Rajan, S.I., and U.S. Mishra. 2011. *The national policy for older persons: Critical issues in implementation*. New Delhi: UNFPA, Working Paper 5.

Selvaraj, S., A. Karan, and S. Madheswaran. 2011. *Elderly workforce participation, wage differentials and contribution to household income*. New Delhi: UNFPA, Working paper 4.

Singh, A., and U. Das. 2012. *Determinants of old age wage labor participation and supply in India: Changes over the past two decades*. http://ssrn.com/abstract=2196183. Accessed on 27 Mar 2013.

UNFPA. 2012. *Report on the status of elderly in selected states of India, 2011*. New Delhi: United Nations Population Fund.

United Nations Population Division. 2013. *World population ageing 2013*. New York: UN, Department of Economic and Social Affairs, Population Division.

Chapter 3
Sample Profile

3.1 Introduction

The data for the study have been collected from two sources. The first source is the large sample surveys on Employment and Unemployment undertaken by the National Sample Survey Office (NSSO) every 5 years. This is a nationally representative survey, containing information on education, age, household expenditure, socio-religious identity, economic status, and labour market participation. Data from the two rounds—55th round (2005–2006) and 68th round (2012–2013)—have been used in the study. Although the NSSO did undertake a survey on Employment and Unemployment in 2009–2010 (62nd Round), the data from this round were not used for three reasons:

(a) There was a drought in the year of survey so that rural income and expenditure were deflated from its normal level.
(b) The quality of data collected in this round has been questioned in internal audits conducted by the Indian Statistical Institute Kolkata.
(c) Five years is too small a period to observe major changes.

Applying an age-based filter, respondents aged 60 years and above were selected. Data for this sub-sample were analysed to obtain a picture of the contribution of aged persons to their families at the national level.

The NSSO survey, however, is not exclusively directed towards aged persons. Therefore, many issues relating to ageing and their contributions to the family remain uncovered in the survey. For instance, we do not have information on extended-SNA (System of National Accounts) activities performed by aged respondents. This is a major deficiency as the aged are mainly engaged in such activities. Nor do we have data on expenditure of aged. The NSSO survey also does not collect information on earnings of persons engaged in the informal sector; this again is a major shortcoming as the aged are mainly engaged in the informal sector.

We have, therefore, supplemented the NSSO data by undertaking a primary survey of aged persons in the capital cities of three states in Eastern India. These states are West Bengal (Kolkata), Orissa (Bhubaneswar), and Jharkhand (Ranchi). The first two states have a high proportion of aged persons (8.5 and 9.5, respectively). In comparison, Bihar and Jharkhand have a low proportion of aged persons (7.4 and 7.1, respectively). We have therefore selected one state (Jharkhand), undertaking the survey in its capital city, Ranchi. The choice was dictated by the high proportion of tribal population in the state. The questionnaire (English version and its Hindi translation) is given in Appendix.

3.2 NSSO Data

The NSS 55th round covered the whole of the Indian Union except (i) Leh and Kargil districts of Jammu and Kashmir, (ii) 768 interior villages of Nagaland situated beyond 5 km of the bus route, and (iii) 172 villages of Andaman and Nicobar Islands, which remain inaccessible throughout the year. A few other areas of Jammu and Kashmir were also excluded from the survey coverage because of unfavourable field conditions.

The survey used the interview method of data collection from a sample of randomly selected households. The fieldwork of 55th round of NSSO started from 1 July 1999 and continued till 30 June 2000. As usual, the survey period of this round was divided into four sub-rounds, each with a duration of 3 months, the first sub-round period covering from July to September 1999, the second sub-round period from October to December 1999, and so on. Equal number of sample FSU (first stage unit) was allotted for survey in each of these four sub-rounds.

The sampling design adopted for the 55th round of NSSO survey was essentially a stratified multi-stage one for both rural and urban areas. The first stage units (FSUs) were villages (panchayat wards for Kerala) for rural areas and NSSO Urban Frame Survey (UFS) blocks for urban areas. The ultimate stage units (USUs) were households. Large FSUs were subdivided into hamlet-groups (rural)/sub-blocks (urban).

The 68th round NSSO survey covered the whole of the Indian Union except (i) interior villages of Nagaland situated beyond 5 km of the bus route and (ii) villages in Andaman and Nicobar Islands, which remained inaccessible throughout the year.

The fieldwork of the 68th round of NSSO started from 1 July 2011 and continued till 30 June 2012. As usual, the survey period of this round was divided into four sub-rounds, each with a duration of 3 months, the first sub-round period ranging from July to September 2011, the second sub-round period from October to December 2011, and so on. An equal number of sample villages/blocks (FSUs) was allotted for survey in each of these four sub-rounds with a view to ensuring uniform spread of sample FSUs over the entire survey period.

A stratified multi-stage design was adopted for the 68th round survey. The first stage units (FSU) were the 2001 census villages (Panchayat wards in case of Kerala)

3.2 NSSO Data 43

in the rural areas and Urban Frame Survey (UFS) blocks in the urban areas. The ultimate stage units (USUs) were households in both the sectors. In case of large FSUs, one intermediate stage of sampling was the selection of two hamlet-groups (hgs)/sub-blocks (sbs) from these FSUs.

Within each district of a State/UT, two basic strata were formed, such as rural stratum comprising all rural areas of the district, and urban stratum comprising all the urban areas of the district. However, within the urban areas of a district, if there were one or more towns with population of 10 lakhs or more as per population census 2001 in a district, each of them formed a separate basic stratum and the remaining urban areas of the district were considered as another basic stratum.

Out of the total number of 12,784 FSUs (7508 villages and 5276 urban blocks) of the central sample allotted for undertaking survey by NSSO, 12,737 FSUs (7469 villages and 5268 urban blocks) could be surveyed at the all-India level for canvassing the Employment and Unemployment schedule (Schedule 10). The number of households surveyed was 1,01,724 (59,700 in rural areas and 42,024 in urban areas).

3.2.1 *Construction of Variables*

From the NSSO survey we used the following information:

(a) *Household characteristics*: Information on household characteristics like household size, household type, religion, social group, land owned, land possessed, land cultivated, and household monthly consumer expenditure was collected.
(b) *Demographic particulars* like age, sex, and educational level were collected.
(c) *Particulars of usual principal activity, subsidiary economic activity, current weekly activity*, and *current daily activity* for all the members of the sample households. For persons engaged in economic activities, information on detailed activity status, industry of work, occupation, wage, and salary earnings by the employees was also collected.

From the NSSO data, we also used the information on the religion (Hinduism, Islam, Christianity, Sikhism, Jainism, Buddhism, Zoroastrianism, and other religions) of the individual and the social group to which the individual belongs (Schedule Tribes, Schedule Caste, Other Backward Class, and other castes). By using religion and social group of the individual, we have constructed the variable SRC (i.e. socio-religious identity of the people). Socio-religious identity of the people is categorized as follows: Muslim, Hindu Upper Caste (HUC), Hindu Schedule Caste (HSC), Hindu Schedule Tribes (HST), Hindu Other Backward Caste (HOBC), and Other Caste (Others).

NSSO data (Schedule 10, section 9) also provide us information of total monthly consumer expenditure. By dividing the total household monthly consumer expenditure by the household size, we arrived at monthly per capita expenditure of the household. In this thesis, for the purpose of analysis, we have considered quintile division of monthly per capita expenditure.

While we have defined labour force, we have taken the self-employed person, regular wage and salary earners, and casual workers and the unemployed. The sub-categories that we have considered in case of self-employed and casual labour are as follows:

Self-Employed

(a) Worked in household enterprises as own-account worker
(b) Worked in household enterprises as an employer
(c) Worked in household enterprises as helper

Casual Labour

(a) Worked as casual wage labour in public works
(b) Worked as casual wage labour in other types of works

In case of work force, we have not considered the unemployed. The persons who have attended educational institution, attended domestic duties, pensioners, not able to work due to disability, and beggars are not included in labour force.

3.2.2 Sample Profile

In the 55th round, data were collected for 592,816 individuals. Within this sample, there were 42,818 persons aged 60 years and above, comprising of 27,326 individuals from rural areas and 15,492 individuals from urban areas. In the 68th round, data were collected for 456,976 individuals, among which there were 38,027 individuals aged 60 years and above. Among the elderly, 23,484 individuals live in rural India and 14,543 in urban India in 2011–2012. Figure 3.1 depicts the percentage of aged people in the population by location of residence and gender. Marginal variation in the break-up of proportion of aged by place of residence and gender is observed.

From Table 3.1 we can see that respondents have low levels of education, with the majority of them being illiterates. The percentage of respondents without education is high in rural areas and among females. Hindu Other Backward Castes (HOBCs) comprise the dominant socio-religious group in rural areas, followed by Hindu Forward Castes (HFCs). In urban areas, on the other hand, HFCs are the majority, followed by HOBCs. More than half of the respondents belong to the age group 60–65 years. Predictably, the population share of older groups is lower.

Table 3.2 gives the sample profile of aged respondents surveyed in the 68th round by NSSO. It can be seen that a large proportion of respondents are without education; this percentage is particularly high among females and in rural areas. HFCs comprise the majority of the respondents; they are followed by HOBCs and Hindu Scheduled Castes (HSCs). About one out of every second person belongs to the young-old group aged 60–65 years. The share of older persons declines as we move to higher age groups.

In Tables 3.3 and 3.4 we have presented the mean of age, household size, and per capita monthly expenditure for the two rounds.

3.2 NSSO Data

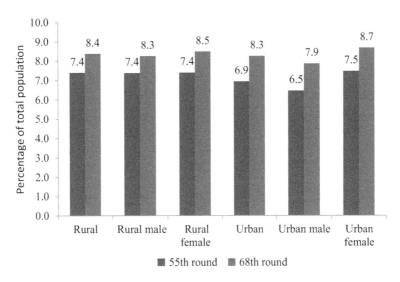

Fig. 3.1 Percentage of aged people in population by place of residence and sex. (Source: Calculated from NSS 68th round)

Table 3.1 Sample profile of aged persons in 55th round of NSSO

Socio-economic correlates		Rural female	Rural male	Urban female	Urban male
Education level	Illiterate	87.53	57.64	60.26	25.98
	Below primary	6.69	17.55	15.24	15.8
	Primary	3.62	10.01	9.52	11.85
	Middle	1.41	6.97	6.84	11.84
	Secondary	0.56	4.91	4.84	15.67
	HS	0.08	1.3	1.29	5.53
	Above HS	0.11	1.62	2.02	13.34
Expenditure group	Poorest	16.45	14.46	17.83	15.08
	Poor	17.54	16.99	19.42	18.29
	Middle	19.48	19.47	20.49	20.17
	Rich	21.65	22.61	20.3	21.79
	Richest	24.89	26.46	21.96	24.66
Socio-religious identity	Muslim	8.76	8.84	13.41	13.59
	HUC	28.49	27.87	42.24	43.16
	HST	6.92	6.70	1.85	1.74
	HSC	14.24	14.68	8.76	8.47
	HOBC	31.7	31.73	23.15	21.7
	Others	9.89	10.18	10.59	11.33
Age group	60–65 years	55	53.59	50.92	51.01
	66–75 years	33.31	34.38	35.63	35.65
	Above 75 years	11.69	12.03	13.44	13.34

Table 3.2 Sample profile of aged persons in 68th round of NSSO

Socio-economic correlates		Rural females	Rural males	Urban females	Urban males
Education level	Illiterate	77.92	43.66	53.12	20.64
	Below primary	9.35	15.09	11.83	10.82
	Primary	6.41	12.11	9.96	10.65
	Middle	3.41	10.14	8.94	12.16
	Secondary	1.77	9.74	7.73	16.84
	HS	0.46	3.9	2.9	7.39
	Above HS	0.69	5.36	5.52	21.51
Expenditure group	Poorest	16.69	16.45	18.87	15.99
	Poor	18.5	18.09	18.6	17.48
	Middle	19.65	18.82	19.37	18.53
	Rich	20.92	21.06	19.37	20.81
	Richest	24.25	25.57	24.03	27.19
Socio-religious identity	Muslim	10.04	11.07	14.62	14.62
	HUC	23.45	22.9	34.46	36.03
	HST	6.57	6.35	2.21	1.68
	HSC	13.97	13.77	10.41	9.37
	HOBC	35.1	35.14	28.43	28.91
	Others	10.87	10.77	9.87	9.39
Age group	60–65 years	54.77	54.32	52.58	52.51
	66–75 years	32.76	33.91	33.3	35.26
	Above 75 years	12.47	11.77	14.12	12.23

Table 3.3 Mean age, household size, and per capita income by gender and place of residence (NSSO, 55th round)

Gender	Rural	Urban	Total
Age (years)			
Male	67.23	67.53	67.33
Female	67.11	67.65	67.31
Total	67.17	67.59	67.32
Household members			
Male	6.39	5.76	6.17
Female	6.16	5.73	6.00
Total	6.28	5.74	6.09
Monthly per capita expenditure (Rs.)			
Male	544.66	890.04	665.13
Female	531.41	835.10	645.28
Total	538.21	861.73	655.26

In the 68th round, marginal variations across gender and place of residence are found for age and household size (Table 3.4). Per capita monthly expenditure is higher in urban areas, representing the higher cost of living and living standards, relative to rural areas.

Table 3.4 Mean age, household size, and per capita income by gender and place of residence (NSSO, 68th round)

Gender	Rural	Urban	Total
Age (years)			
Male	67.00	67.00	67.00
Female	67.00	68.00	68.00
Total	67.00	68.00	67.00
Household members			
Male	6.00	5.00	5.00
Female	5.00	5.00	5.00
Total	6.00	5.00	5.00
Monthly per capita expenditure (Rs.)			
Male	1573.75	2570.19	1948.81
Female	1518.05	2403.41	1861.94
Total	1546.06	2485.15	1905.20

3.3 Primary Survey

As mentioned before, the primary survey was undertaken by us in three states. Field investigators were students of different courses, with some field experience. They were personally trained through interactive sessions. In addition, close monitoring of the questionnaires submitted initially was undertaken to identify errors or misunderstandings, or resolve doubts. The survey was undertaken between November 2016 and January 2017 in Bhubaneswar, February 2017 and April 2017 in Kolkata, and May 2017 and July 2017 in Ranchi. About 10% wards were randomly selected in each city, and survey administered to permanent residents aged 60 years or more residing with their families. After cleaning and editing, a total of 518 questionnaires were retained. The gender and city break-up are given in Fig. 3.2.

Table 3.5 presents the socio-economic characteristics of the sample. Given Kolkata's large size, we had opted for a larger sample size in Kolkata, relative to the other two cities. About half of the respondents are aged 60–65 years; the majority of respondents belong to the Hindu community and fall under the General caste category. If we consider household type, we will find that most of the respondents come from the residual 'Others' category, followed by regular wage and salaried earning group. Analysis of the educational profile reveals that persons with no education, 6–10 years of schooling, and graduates and above form the major chunk of respondents. About 65% of respondents reside in families with 2–5 members.

In Table 3.6 we have given the average age, household size, and per capita monthly expenditure by gender and city. While little variations across gender and cities are observed in terms of mean age, respondents from Kolkata have smaller size families and have higher expenditure levels than the other two cities. This is to be expected, given that Kolkata is a A-class metropolitan city.

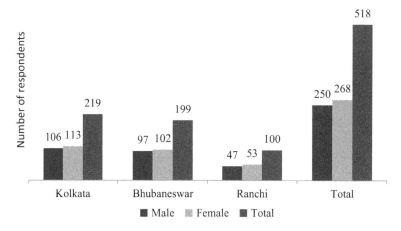

Fig. 3.2 Number of respondents by gender and place of residence

Table 3.5 Sample profile of respondents of primary survey

Correlates	Male	Female	Total
City			
Kolkata	42.4	42.16	42.28
Bhubaneswar	38.8	38.06	38.42
Ranchi	18.8	19.78	19.31
Age groups			
60–65 years	50.4	47.76	49.03
66–70 years	19.2	20.52	19.88
71–80 years	22.00	21.27	21.62
81 years and above	8.40	10.45	9.46
Religion			
Hinduism	92.40	82.09	87.07
Islam	2.40	7.84	5.21
Christianity	3.20	5.60	4.44
Others	2.00	4.48	3.28
Caste			
Scheduled Tribe (ST)	8.00	9.70	8.88
Scheduled Caste (SC)	5.20	11.94	8.69
Other Backward Castes (OBC)	18.40	20.52	19.50
General	68.40	57.84	62.93
Household type			
Self-employed	18.40	13.81	16.02
Regular wage earner	20.40	29.48	25.10
Casual labour	10.00	5.22	7.53
Others	51.20	51.49	51.35

(continued)

3.3 Primary Survey

Table 3.5 (continued)

Correlates	Male	Female	Total
Education			
No education	12.40	36.94	25.10
Primary or below	14.40	13.43	13.9
Middle and secondary	29.60	21.27	25.29
Higher secondary	15.20	7.84	11.39
Above high school	28.40	20.52	24.32
Household size			
2–3 members	25.61	21.35	23.39
4 members	22.76	20.22	21.44
5 members	17.07	23.97	20.66
6 members	11.79	12.36	12.09
7–10 members	17.07	17.23	17.15
Above 10 members	5.69	4.87	5.26

Table 3.6 Mean age, household size, and per capita income by gender and city

City	Male	Female	Total
Age (years)			
Kolkata	68.51	68.22	68.36
Bhubaneswar	68.23	68.17	68.20
Ranchi	68.57	70.32	69.50
Total	68.41	68.62	68.52
Household members			
Kolkata	3.91	4.12	4.02
Bhubaneswar	6.04	6.15	6.10
Ranchi	7.09	6.81	6.94
Total	5.30	5.42	5.36
Per capita expenditure monthly (Rs.)			
Kolkata	6263.06	6223.97	6242.89
Bhubaneswar	5695.20	3102.25	4366.15
Ranchi	5276.96	5636.17	5473.58
Total	5866.78	4916.93	5372.41

We now turn to analysis of our results.

Chapter 4
Economic Contribution of the Aged: A National Profile

4.1 Introduction

The starting premise of this study was that, as life cycles are prolonged, they create challenges for society and the state. With the inversion of the age pyramid, the dependency ratio increases sharply, placing greater pressure on the younger generations. Simultaneously, in many Asian countries, the changing nature of society has led to the disintegration of traditional structures that provided socio-economic security to the aged, thereby increasing their vulnerability. In an earlier chapter, we argued that such social and demographic changes may generate a pressure on the aged to re-enter the labour market after their 'retirement'. Chapter 2 had briefly summarized different facets of this trend—the extent to which the aged are able to rejoin the workforce in an increasingly competitive labour market, the occupational structure and quality of employment, and finally the earnings of the aged. Postponement of retirement by the aged, however, has another important implication for the household.

The literature on gerontology has typically viewed the aged as a burden on the state and society. While delivering a special lecture at the European Population Conference in June 2014, Paul Demeny referred to the direct and indirect burdens imposed by ageing on the young. When the elderly work, however, they are able to make a financial contribution to the household. This contribution can sometimes be quite substantial. Even if it is numerically a small proportion of total household income, such contribution can sometimes play a critical role by ensuring that the family tides over a temporary economic crisis. Psychologically, too, the contribution enables the aged to live with dignity, as the label of an 'economic burden' no longer sticks to them.

In this chapter, we examine this aspect. Our analysis focuses on the monthly income earned by aged members of the family as a percentage of household expenditure. Variations across socio-economic status and over regions are examined. We also estimate a multiple linear regression model (MLRM) to identify determinants of contribution of aged members to household expenditure.

4.2 Methodology

We have seen that information on earnings of workers and number of days in the week preceding the survey is provided in Block 5.3 of the National Sample Survey Office (NSSO) questionnaire (relating to current weekly status). Using these data, it is possible to calculate daily earnings of workers. Daily earnings are multiplied by 30 to arrive at monthly earnings. However, NSSO does not provide the data related to earnings of informal sector workers either as own account workers or as workers in household enterprises. Not considering earnings of the elderly in the informal sector underestimates their true contribution as the informal sector is the largest employment providing sector, both in general and to the aged workers, in India. Thus, in this chapter, we attempt to estimate the earnings of aged workers from the informal sector.

Given that

$$\text{Monthly earnings} = \text{Monthly expenditure} + \text{Savings}$$
$$= \text{Income from workers engaged in formal sector}$$
$$+ \text{Income from workers engaged in informal sector}$$

If savings are assumed to be 0, a not unreasonable assumption for households who are dependent on income from informal sector, we can estimate income from informal sector as

$$\text{Income from workers engaged in informal sector}$$
$$= \text{Monthly expenditure} - \text{Income from workers engaged in formal sector}$$

We should keep in mind that the income from informal sector is contributed by both the elderly and non-elderly workers in the household. To obtain the income of aged workers working in the informal sector, we assume that both elderly and non-elderly workers secure work for the same number of days. Dividing the informal sector earnings with the total workers in the informal sector, we arrive at average monthly earnings of informal sector workers in the households. If this figure is multiplied by the number of elderly workers in the informal sector in each household, we get informal sector earnings of aged workers in such households.

The aggregate monthly earnings of all aged workers (including earnings of informal sector workers) in each household are used to estimate the gross financial contribution of the aged. This is defined as follows:

4.2 Methodology

$$\text{Gross financial contribution of aged} = 100 \times \frac{\left(\text{Total monthly earnings of all aged workers}\right)}{\text{Total monthly household expenditure}}$$

Using the above formula, we estimate mean gross contribution in rural and urban areas, and in both the rounds. The above formula gives some idea of the financial contribution of the elderly to the household. However, a more appropriate indicator to capture the actual contribution of the elderly to the household should be one that deducts, from gross contribution, the amount of money income spent for their (elderly) own expenses. Thus, we have focused on the net contribution of the elderly, given as follows:

$$\text{Net financial contribution of aged} = 100 \times \frac{\left(\text{Total monthly earnings of all aged workers} - \text{Total monthly expenditure on aged}\right)}{\text{Total monthly household expenditure}}$$

Net contribution helps us to analyse whether the elderly are contributing more to the family than their share of consumption expenditure.

Unfortunately, NSSO does not provide information on member-wise expenditure; it provides aggregate expenditure for the family as a whole. Although expenditure on some items, for instance, healthcare expenditure, may be higher for the elderly, per capita expenditure of aged members may not differ significantly from per capita expenditure of non-aged members as expenditure on other components, such as food, transport, and recreation, may be lower for the aged. Dividing monthly expenditure with the total household members, we arrive at a total monthly per capita expenditure of the household.

Mean contribution (net) of the aged is analysed for different expenditure classes. State-wise variations in the mean contribution (net) are analysed for both the rounds. We then estimated a t-test to check whether these contributions are statistically significant for all the states. The spatial variation in the mean net financial contribution of the elderly workers in different states is presented for both rural and urban areas in both the rounds to obtain an idea of the geographical distribution and pattern of the financial contribution of aged workers. The mean net contribution is then compared with the proportion of aged in state population (obtained from the 2001[1] and 2011 Censuses[2]) and the per capita net state domestic product (NSDP, reported by Directorate of Economics and Statistics for state governments[3]). The next part of the

---

[1] Accessed from www.censusindia.gov.in/2001census/c9_India.pdf on 01/05/2014.
[2] Accessed from www.censusindia.gov.in/2011census/populationenumeration.aspx on 03/05/2014.
[3] Accessed from http://mospi.nic.in/Mospi_New/upload/statewise_sdp1999_2000_8feb.pdf on 10/05/2014.
  Accessed from http://mospi.nic.in/Mospi_New/upload/State_wise_SDP_2004-05_14mar12.

analysis is based on an econometric model that identifies determinants of the net financial contribution of the elderly workers.

In the second part of our analysis, we will analyse the effect of the financial contribution of elderly on household poverty. As monthly per capita expenditure (MPCE) is household specific, we have first estimated household-level poverty in different states in India, and once again household poverty in different states is estimated excluding the financial contribution of the aged from household expenditure with respect to the number of non-aged members in the household in both the rounds, i.e. household poverty is again estimated by using the following variable MPCE'.

$$\text{MPCE}' = \frac{\left(\text{Total monthly household expenditure} - \text{Total monthly earnings of all aged workers}\right)}{\text{Number of non aged member in the household}}$$

Our objective is to find out whether the contribution of aged persons helps to reduce household poverty significantly or not and whether there is any change in the poverty situation over the study period. The Planning Commission's estimate of state-specific poverty line for the year 1999–2000[4] and 2011–2012[5] has been used for the purpose of analysis. To identify the proportion of poor in the population, the most widely used measure, head count ratio (HCR), has been used here.

$$\text{HCR} = q/n$$

where q = Number of poor, n = Total number of persons.

Although HCR is widely used, it has some limitations, e.g. HCR is insensitive to income transfer within the group of poor. Another difficulty with HCR is that the poor could be just below or far below the line, but HCR would not register any distinction between these two situations. Keeping in mind these limitations, we have focused on the Foster, Greer, and Thorbecke (FGT) index, which captures both the level and intensity of poverty (Foster et al. 1984). The FGT index is given by

$$\text{FGT} = 1/N \sum (G_i / Z)^\alpha$$

where N = Total number of persons
Z = Poverty line
G_i = Income shortfall of the ith individual, and $\alpha \geq 0$

For different values of α, we get different indices. As the index, when $\alpha = 2$ (FGT2), satisfies all the axioms of poverty index, such as continuity, invariance condition, monotonicity, strong monotonicity, transfer axioms, and decomposability, we use FGT2. This chapter estimates both the HCR and FGT2 for different

---

pdf on 02/05/2014.

[4] Accessed from http://planningcommission.nic.in/pov_rep0707.pdf on 15/01/15.

[5] Accessed from http://planningcommission.nic.in/pre_pov.pdf2307 on 18/01/15.

4.2 Methodology

states in India in the 55th and 68th rounds of NSS. To check whether the contribution of the elderly helps to reduce household poverty significantly or not, we have estimated 't' statistics corresponding to different states in both the rounds of NSS using the HCR. Changes in household poverty due to the contribution of elderly in different states are also compared with the proportion of elderly, state domestic product, etc. in each round, and such analysis is graphically presented in this chapter.

The functional specification and econometric model regarding the net financial contribution of the elderly are described in the following sub-sections.

4.2.1 Functional Specification

As in earlier chapters, in this chapter, we have determined whether there has been a temporal change in the net financial contribution of the aged by regressing net contribution of elderly to household expenditure (NCONT) on a Time dummy in a pooled model of both rounds. Control variables are also included in the model. This model is reported in Appendix. Subsequently, the determinants of financial contribution have been analysed disaggregating the sample by place of residence across rounds. This model is of the form:

$$NCONT = \alpha + \beta_2 LPCME + \beta_3 LPCME^2 + \beta_4 HHSIZE + \beta_5 LNSDP + \beta_6 NON\ AGED\ WORKING\ MEMBER + \beta_7 MUSLIM + \beta_8 HST + \beta_9 HSC + \beta_{10} HOBC + \beta_{11} HOTHERS + \beta_{12} SENA + \beta_{13} AGLAB + \beta_{14} RLAB + \beta_{15} ROTHERS + \beta_{16} USE + \beta_{17} CASUAL + \beta_{18} UOTHERS + \beta_{19} UNEMP$$

(4.1)

where

NCONT = Net contribution of elderly to household expenditure
LPCME = Log of monthly per capita expenditure
HHSIZE = Size of household of the respondent
LNSDP = Log of per capita net state domestic product
UNEMP = State-level unemployment
NON-AGED WORKING MEMBER = Number of working members aged 15–59 years in the family
MUSLIM = 1 if the respondent is a Muslim, = 0 otherwise
HSC = 1 if the respondent is a Hindu Scheduled Caste, = 0 otherwise
HST = 1 if the respondent is a Hindu Scheduled Tribe, = 0 otherwise
HOBC =1 if the respondent is a Hindu Backward Caste, = 0 otherwise
HOTHERS = 1 if the respondent belongs to all others socio-religious identity, = 0 otherwise
(HUC, i.e. Hindu Forward Caste, is the reference category)
SENA = 1 if the respondent is self-employed in non-agriculture, = 0 otherwise

AGLAB = 1 if the respondent works as agricultural labour, = 0 otherwise
RLAB = 1 if the respondent works as rural labour, = 0 otherwise
ROTHERS = 1 if the respondent works in all others jobs in rural area, = 0 otherwise
(SEAGRI, i.e. self-employed in agriculture, is the reference category)
USE = 1 if the respondent is self-employed in urban area, = 0 otherwise
CASUAL = 1 if the respondent works as casual labour, = 0 otherwise
UOTHERS = 1 if the respondent works in all other jobs in urban area, = 0 otherwise
(WSA, i.e. urban wage and salary earners, is the reference category)
State-specific fixed effects have also been incorporated into the model

The equation is a multiple linear regression model (MLRM) so that the parameters are estimated by using ordinary least square (OLS) method.

4.3 Results and Discussion

4.3.1 Financial Contribution of Elderly to Household Expenditure

In a country like India, where most of the people work outside the domain of the formal sector, people hardly get any pension benefit during their old age. Nowadays, as already discussed, the rising cost of living and increasing healthcare cost force the elderly to participate in the work and, thereby, provide some support to the family members by contributing to the household's economic status.

The analysis reveals (Fig. 4.1) that in the 55th and 68th rounds, in about 71% of the household, the net financial contribution of the aged is negative. In such households, aged members are dependent on other household members. This implies that, in India, the elderly consume more than what they earn in the majority of households. It has also been observed that almost in 28% of the households in the 55th round and 27% of the households in the 68th round, the elderly are 'assets' as they are making a positive net contribution to their families.

Across rounds, the percentage of households in which elderly are a 'burden' or an 'asset' almost remains the same in both rural and urban India (Fig. 4.2). In rural India, elderly are 'assets' in 32% of households in both the rounds; in urban India, the corresponding percentage is about 20% in both rounds. In urban India, net contribution of the aged is negative in about 80% of households. It is slightly lower in rural households.

In the next part, we have analysed the gross and net financial contributions of the elderly people. On an average, the gross contribution of elderly workers has been found to increase from 19% to 21% of the household expenditure at the all-India level over the study period (Fig. 4.3). Both rural and urban India has experienced an increase in the gross contribution of the aged to household expenditure over the

4.3 Results and Discussion

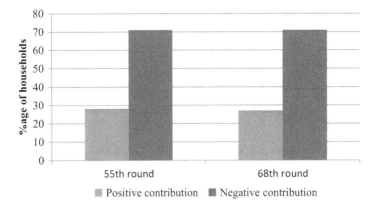

Fig. 4.1 Percentage of households where elderly are burden or asset in India in the 55th and 68th rounds of NSS. (Source: Calculated from NSS 55th and 68th rounds)

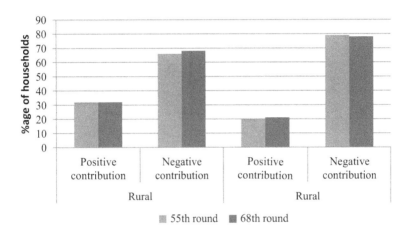

Fig. 4.2 Distribution of households by net contribution of aged members in rural and urban India in the 55th and 68th rounds of NSS. (Source: Calculated from NSS 55th and 68th rounds)

study period. However, the gross contribution of elderly in rural India is much higher than in urban India. Furthermore, in recent years, the increase in gross financial contribution is higher among the rural elderly households compared to urban elderly households. This may be because of higher work participation of the elderly in rural areas.

However, as already mentioned, net contribution—which takes into account the expenditure incurred on the elderly—is a more appropriate indication of actual contribution to the family. Our analysis reveals that the net financial contribution of the elderly people is negative at the all-India level; this also holds for both rural and urban India (Fig. 4.4). It implies that the monthly expenditure of the elderly is higher than the income that they earn. However, this figure should not be taken to

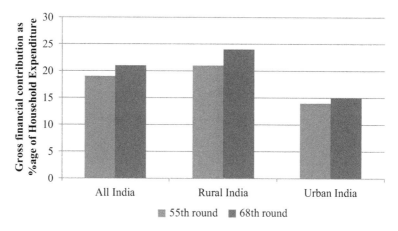

Fig. 4.3 Gross contribution of the elderly to household expenditure in the 55th and 68th rounds of NSS. (Source: Calculated from NSS 55th and 68th rounds)

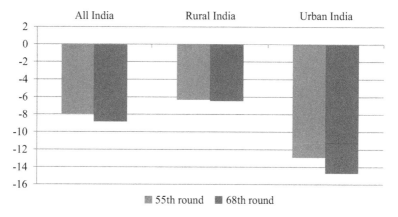

Fig. 4.4 Net contribution of the elderly to household expenditure in the 55th and 68th rounds of NSS. (Source: Calculated from NSS 55th and 68th rounds)

indicate that the aged are a burden on their families. First, workers may have saved during their working career, so that they are dis-saving after retirement. Second, particularly in urban areas, female members may participate in the labour market. In such cases, aged members may contribute to family welfare in non-financial ways by caring for children, going to shops, etc. facilitating participation of female members in the labour market.

The negative contribution is much higher in absolute terms in the 68th round compared to the 55th round. The more negative contribution may be because of work participation rate of the elderly has decreased in recent years, and there is a greater degree of informalization in India. The difference between rural and urban India regarding net contribution of the elderly may reflect the fact that the elderly are participating more in the workforce in rural India compared to urban India (Selvaraj et al. 2011) and, therefore, contributing more to the household expenditure, as the contribution of the elderly is linked to the current work status (UNFPA 2012). Another possible reason is that affluent people generally live in urban areas, and they are more aware; thus, they may have saved more during their young age and they are dis-saving in their old age. Apart from that, the pension earnings are high in urban India. Thus, the elderly people may meet their monthly expenses out of savings and pension.

4.3.2 Changes in Net Financial Contribution Across Socio-economic Strata

Analysis of the financial contribution of the elderly across economic strata is extremely important. Table 4.1 presents estimates of average net financial contribution by quintile divisions of monthly per capita expenditure. A negative relation between the monthly net financial contribution of the elderly to household economic status and quintile division of monthly per capita expenditure groups in India in the 55th round and 68th round of NSS has been observed at the all-India level. Similar results have been found in rural and urban India in both the rounds.

In both rural and urban India, a sharp decline in contribution across expenditure quintile classes is observed. In rural and urban India, in both the rounds, the elderly respondents from the poorest quintile class are found to be contributing more than the aged from more affluent households. The relatively high contribution of the lowest expenditure group of elderly implies that economic pressure and need for survival is a major driving force behind participation in the labour market (Auer and Fortuny 2000; Milbourne and Doheny 2012). A relatively lower contribution of the elderly belonging to the higher expenditure quintile group may be because of the

Table 4.1 Net financial contribution of the elderly workers by expenditure group and location of residence in 55th and 68th rounds of NSS (percentage)

Place of Residence	Round	Expenditure Group				
		Poorest	Poor	Middle	Rich	Richest
All India	NSS 55th	−6.78	−6.51	−6.60	−8.01	−13.59
	NSS 68th	−6.1	−6.8	−7.2	−11.3	−16.7
	Difference	**0.69**	**−0.32**	**−0.55**	**−3.24**	**−3.09**
Rural India	NSS 55th	−6.22	−6.70	−5.82	−6.26	−6.67
	NSS 68th	−5.0	−6.8	−6.4	−8.2	−6.9
	Difference	**1.17**	**−0.05**	**−0.54**	**−1.98**	**−0.20**
Urban India	NSS 55th	−8.52	−9.02	−12.23	−15.38	−19.84
	NSS 68th	−9.0	−9.7	−13.9	−18.2	−21.6
	Difference	**−0.44**	**−0.70**	**−1.68**	**−2.77**	**−1.73**

Source: Calculated from NSS 55th and 68th rounds

better-off economic condition of the younger household members, or because the aged have saved during their working period. This reduces pressure to contribute to their families by prolonging retirement. Furthermore, the work opportunities of the affluent aged are more restricted as they will work only in certain conditions; poor aged, on the other hand, do not have the luxury to pick and choose.

Over the study period, at the all-India level, net contribution has increased only among the poorest. Rural India also has registered the same scenario. On the other hand, urban India has experienced a decrease in the contribution of the elderly for all expenditure groups in 2011–2012 compared to 1999–2000.

4.3.3 Regional Analysis of Financial Contribution of the Elderly

In the next part of our analysis, we have disaggregated the results at the regional level. We examine regional variations in contribution of the elderly to their families over the study period using choropleth maps. The net contributions of the elderly people in different states of rural and urban India in both the rounds are shown in the Appendix Table (Table 4.5). The estimated '*t*'-statistics (Table 4.6) confirms that, almost in all the states, the elderly are significantly contributing to household expenditure in both the rounds in rural India and in urban India.

4.3.3.1 Regional Analysis of Financial Contribution in the 55th Round of NSS

Analysis of the financial contribution of the elderly across different states in the 55th round (Fig. 4.5) in rural India shows that the contribution of the elderly to household economic status is negative in most of the states (like Rajasthan, Madhya Pradesh, Gujarat, Punjab, West Bengal, Orissa, Kerala, Karnataka, Andhra Pradesh, etc.). A positive net financial contribution is observed in only a few cases (like rural Uttar Pradesh, Chandigarh, etc.).

The contribution of the elderly is more marked in the rural North and North-East India—in Jammu and Kashmir, Himachal Pradesh, Chandigarh, Uttar Pradesh, Manipur, Nagaland, etc. In extreme South Indian states such as in rural areas of Kerala, Karnataka, Andhra Pradesh, Tamil Nadu, and Maharashtra, the contribution of the elderly is less marked in the 55th round.

In the next part, we have examined the average financial contribution of the elderly in urban India in the 55th round of NSS. Figure 4.6 depicts that almost in all the states of urban India, the contribution of the elderly people is negative. Average financial contribution of urban elderly is high in states such as Bihar, Uttar Pradesh, Haryana, Delhi, and Madhya Pradesh. On the other hand, net financial contribution

4.3 Results and Discussion

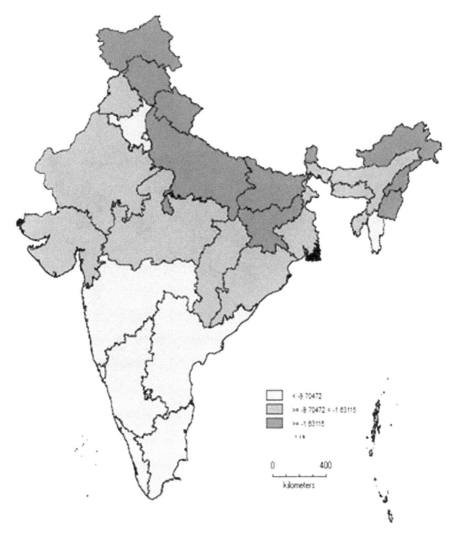

Fig. 4.5 Financial contribution of the elderly in different states of rural India in the 55th round of NSS (percentage). (Source: Calculated from NSS 55th round)

is low in southern states such as Tamil Nadu, Andhra Pradesh, and Karnataka and in West Bengal, Orissa, and Gujarat.

A question that arises in this context is what are the state-level factors that cause variations in the magnitude of the financial contribution of aged. Of particular interest is the impact of economic growth (creating employment opportunities for the aged) and age structure of the population (a high share of the elderly should

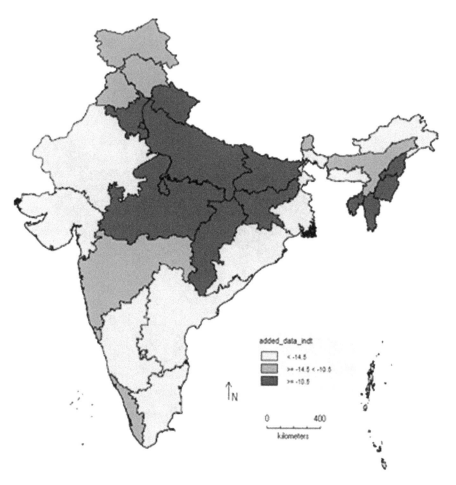

Fig. 4.6 Financial contribution of the elderly in different states of urban India in the 55th round of NSS (percentage). (Source: Calculated from NSS 55th round)

increase the old dependency ratio[6] and create a pressure on the aged to work) (James 1994, Subaiya and Bansod 2011).

In order to examine whether interstate variations in contribution of elderly towards household's economic status can be explained by the differences in the proportion of elderly or by the economic growth in the states, we have plotted the scatter of mean contribution in different states against the proportion of elderly, net state domestic product, etc., in both the rounds for rural and urban India.

According to the 2001 Census, the proportion of rural elderly was high (lies between 8% and 11%) in the states such as Andhra Pradesh, Himachal Pradesh,

---

[6] Dependency ratio is the ratio of dependents (children and aged) to total population. This may be split into young dependency ratio and old dependency ratio, with the numerator being adjusted accordingly.

4.3 Results and Discussion

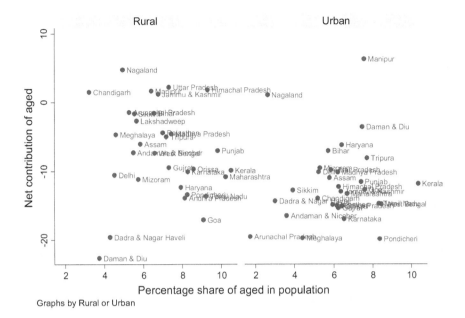

Fig. 4.7 State-wise relationship between average financial contribution and proportion of elderly in rural and urban India in the 55th round of NSS (percentage). (Source: NSS 55th round and Census 2001)

Karnataka, Kerala, Maharashtra, Orissa, Punjab, and Tamil Nadu, and in urban India, the same has been found in the states such as Kerala, Tamil Nadu, and West Bengal (Table 4.7). From Fig. 4.7, a positive relation between the proportion of the elderly in the states and positive net contribution towards household's economic status is observed in urban India. In rural India, no clear pattern is observed between the study variables.

With respect to the log of net state domestic product per capita in rural and urban India in the 55th round, no clear pattern is observed (Fig. 4.8).

4.3.3.2 Regional Analysis of Financial Contribution in the 68th Round of NSS

Analysis of the graph (Fig. 4.9) for rural India in the 68th round reveals that net contribution is negative in most of the states such as Gujarat, Rajasthan, Maharashtra, Andhra Pradesh, Kerala, and Tamil Nadu. However, positive contribution of the elderly has been observed in rural Uttar Pradesh, Chandigarh, and Jharkhand.

The financial contribution of aged workers in 2011–2012 is high in Central India such as rural Uttar Pradesh, Jharkhand, Madhya Pradesh, Chhattisgarh, Chandigarh,

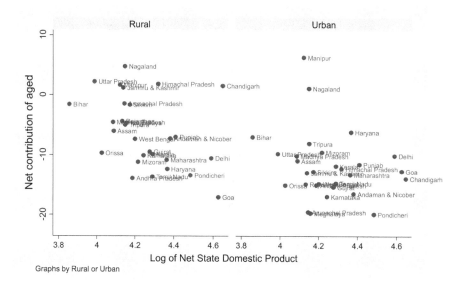

Fig. 4.8 State-wise relationship between the average financial contribution and log of net state domestic product in rural and urban India in the 55th round of NSS (percentage). (Source: NSS 55th round)

Rajasthan, and also in some northeast states such as Meghalaya, Nagaland, and Manipur. However, in southern India (Kerala, Tamil Nadu, and Andhra Pradesh) and in rural West Bengal, Orissa, Maharashtra, and Goa, the average net financial contribution is low in the 68th round.

In 2011–2012, the contribution of elderly workers residing in urban areas does not indicate a clear spatial pattern (Fig. 4.10). In all the states, contribution of the aged to their families is negative. The average net contribution is high in Uttar Pradesh, Bihar, Rajasthan, Karnataka, and in the North-East Indian states such as Manipur, Mizoram, and Meghalaya. Lowest contribution of the aged has been observed in Gujarat, West Bengal, Maharashtra, and Andhra Pradesh.

Graphical presentation of the relationship between the proportion of the elderly in the population and net contribution of the elderly in different states of rural and urban India in the 68th round is illustrated in Fig. 4.11. The proportion of the elderly with respect to population in rural India is quite high in all the states of South India and in rural Punjab (more than 10%), and in urban India, the same has been found in Kerala, Tamil Nadu, and West Bengal (Table 4.7) in recent years. From Fig. 4.11, we do not find any clear relation between the net contribution of the elderly and proportion of the elderly in the states.

Other than the percentage of elderly, the pace of economic growth is one of the important determinants of the financial contribution of elderly. Figure 4.12 shows a negative relationship between the net financial contribution of elderly and log of net

Fig. 4.9 Financial contribution of the elderly in different states of rural India in the 68th round of NSS (percentage). (Source: Calculated from NSS 68th round)

state domestic product in rural India in the 68th round of NSS. However, in urban India, no clear pattern is found between the study variables.

4.4 Determinants of Financial Contribution of Elderly

In the following part, we will analyse the effects of determinants on the net financial contribution of the elderly at the household level in both the rounds. The analysis will be undertaken for rural and urban areas, separately.

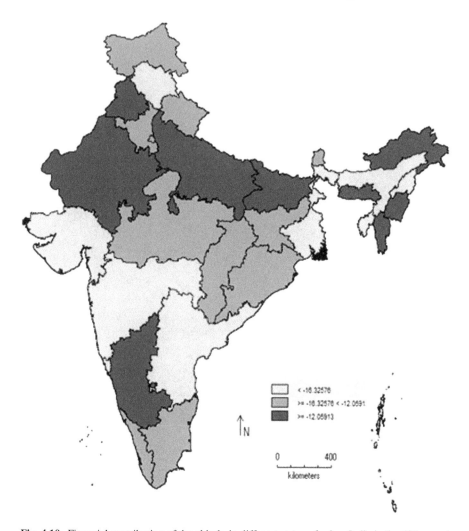

Fig. 4.10 Financial contribution of the elderly in different states of urban India in the 68th round of NSS (percentage). (Source: Calculated from NSS 68th round)

4.4.1 Econometric Analysis

Results of the econometric analysis for the rural sub-sample reveal that the F-statistics are statistically significant at 1% level in both the rounds (Table 4.2). This implies that the null hypotheses H_0: $(\beta_1 = \beta_2 = \ldots = \beta_k)$ is rejected. The R^2 statistics shows that the model explains 15% and 22% of the variation in dependent variable in the 55th and 68th rounds, respectively.

The result, reported in Table 4.2, reveals that the household expenditure level does not have any significant relation with net contribution of the rural elderly in both the

4.4 Determinants of Financial Contribution of Elderly 67

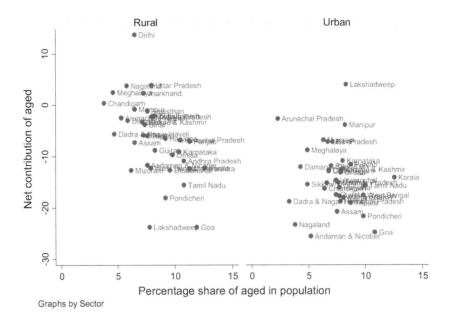

Fig. 4.11 State-wise relationship between the average financial contribution and proportion of the elderly in rural and urban India in the 68th round of NSS (percentage). (Source: NSS 68th round)

rounds. Predictably, rural elderly workers contribute more in larger families and in families with a smaller number of non-aged working members in both the rounds. State-level unemployment is negatively associated with net financial contribution of the elderly in both the rounds. In rural India, per capita net state domestic product does not show any significant relationship with the net contribution of elderly in the 55th round of NSS. However, in the 68th round, we observe a negative relationship between the log of net state domestic product and net contribution of elderly.

Significant variations in the magnitude of net financial contribution are observed across socio-religious groups in the 55th and the 68th rounds in rural India. In the 55th round, we found that the contribution of aged is relatively higher in Hindu Scheduled Tribes households, compared to that in Hindu Forward Caste households. In the 68th round, the aged from not only Hindu Scheduled Tribe households but also from Muslim, Hindu Scheduled Caste, and Hindu Other Backward Caste households contribute more than the aged in Hindu Upper Caste families.

The contribution of the aged members who are self-employed in the non-agricultural household does not significantly differ from the elderly who are self-employed in agriculture in the 55th round. However, in the 68th round, their net financial contribution is significantly higher than self-employed elderly in agriculture. Although, an agricultural labourer, rural labourer household, elderly contribute less to household expenditure compared to farming households in the 55th round, they are found to be contributing more in the 68th round.

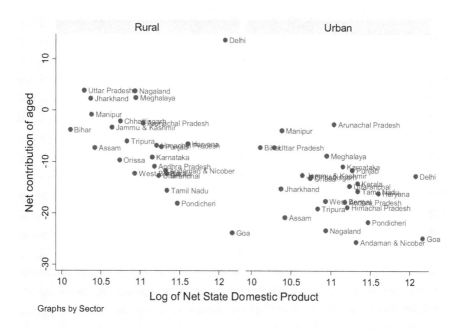

Fig. 4.12 State-wise relationship between the average financial contribution and log of net state domestic product in rural and urban India in the 68th round of NSS (percentage). (Source: NSS 68th round)

Table 4.3 reports regression results for net financial contribution of the elderly for the urban sub-sample. The F-statistic is significant at 1% level; furthermore, the independent variables explain 16% and 18% variations in the dependent variables in the 55th and 68th rounds of NSS.

Similar to the results for rural India, in urban India, in both the rounds, household monthly per capita expenditure does not display any significant relationship with a net contribution of the elderly. In urban India, the elderly are contributing more in larger families and in families with a smaller number of non-aged working members. This is observed in both the rounds. Per capita net state domestic product shows a significant (and negative) relationship with net financial contribution only in the 68th round. State-level unemployment shows significant and negative relation in both the rounds.

Among different socio-religious group, the elderly from Muslim and Hindu Other Backward Caste households are contributing significantly more to their families than the elderly from Hindu Upper Caste; this is observed for both rounds. Aged members from Hindu Scheduled Tribes are contributing significantly more than the Hindu Upper Caste elderly; this is true in only the 68th round.

In an urban area, the elderly workers from self-employed households are contributing significantly more to household's economic status compared to wage and salary earning households in both the rounds. However, the contribution of the elderly

4.5 Contribution of the Aged and Poverty Levels

Table 4.2 Effects of predictor variables on net financial contribution of elderly in rural India in the 55th and 68th rounds of NSS

	NSS 55th		NSS 68th	
Variables	coefficient	Prob.	coefficient	Prob.
LMPCE	0.70	0.92	−8.52	0.19
LMPCE2	−0.03	0.95	0.56	0.21
HH SIZE	2.44	0.00	2.82	0.00
LNSDPC	−2.72	0.52	−2.88	0.07
NON-AGED WORKING MEMBER	−6.33	0.00	−9.27	0.00
UNEMP	−52.49	0.00	−45.06	0.00
SOCIO-RELIGIOUS IDENTITY (RC: HINDU UPPER CLASS)				
MUSLIM	1.27	0.12	4.20	0.00
HST	3.99	0.00	2.57	0.02
HSC	−0.22	0.76	1.56	0.07
HOBC	0.54	0.34	1.33	0.04
HOTHERS	0.005	1.00	−0.06	0.96
HOUSEHOLD TYPE (RC: SELF-EMPLOYED IN AGRICULTURE)				
SENA	−0.87	0.17	24.51	0.00
AGLAB	−15.68	0.00	19.08	0.00
RLAB	−10.62	0.00	6.80	0.00
ROTHERS	−18.36	0.00	−17.49	0.00
STATE DUMMIES	YES		YES	
INTERCEPT	21.31	0.63	38.0135	0.224
N	18,917		13,936	
F	79.06	0.00	104.93	0.00
R^2	0.15		0.22	

Source: Calculated from NSS 55th and 68th rounds

workers from casual labourer households is not significantly different from that of the aged from wage and salary earning households.

4.5 Contribution of the Aged and Poverty Levels

Nowadays, poverty among the aged population is a major concern in many countries of the world (Barrientos et al. 2003; Srivastava and Mohanty 2012). The poverty rate of the Korean elderly in 2000 was three times higher than the non-elderly partly because of inadequate government pension support in spite of the rapid economic growth of the country (Park et al. 2003). A possible reason for this is reduction in activity levels, along with declining health condition with age, which may lead to high expenditure on health, increasing the risk of poverty. Furthermore, less-educated elderly, those who are living alone or in a rural area or in a poor health condition, are not benefitting from poverty reduction policies; this has an adverse effect on economic well-being of the elderly (Jinkook and Drystan 2011).

Table 4.3 Effects of predictor variables on net financial contribution of elderly in urban India in the 55th and 68th rounds of NSS

Variables	NSS 55th Coefficient	Prob.	NSS 68th Coefficient	Prob.
LMPCE	2.41	0.67	−8.31	0.25
LMPCE2	−0.39	0.36	0.39	0.41
HH SIZE	2.34	0.00	2.96	0.00
LNSDPC	1.75	0.48	−7.68	0.00
NON-AGED WORKING MEMBER	−5.85	0.00	−8.85	0.00
UNEMP	−41.58	0.00	−38.31	0.00
SOCIO-RELIGIOUS IDENTITY (RC: HINDU UPPER CLASS)				
MUSLIM	1.79	0.03	3.07	0.00
HST	0.89	0.64	3.49	0.10
HSC	2.86	0.00	0.58	0.58
HOBC	2.00	0.01	2.31	0.00
HOTHERS	0.18	0.85	2.75	0.01
HOUSEHOLD TYPE (RC: WAGE and SALARY EARNERS)				
USE	10.08	0.00	9.65	0.00
CASUAL	0.65	0.50	1.66	0.11
UOTHERS	−16.99	0.00	−26.26	0.00
STATE DUMMIES	YES		YES	
INTERCEPT	−33.20	0.26	109.06	0.00
N	10,969		9604	
F	49.43	0.00	59.49	0.00
R^2	0.1564		0.18	

Source: Calculated from NSS 55th and 68th rounds

However, the rates of poverty differ among the countries. Jinkook and Drystan (2011) have found that although the poverty rate among the elderly has decreased significantly in Korea in 2008 compared to 2006, the rate is higher than other Organisation for Economic Co-operation and Development (OECD) countries. Morris (2007) reported that robust economic growth and government assistance programmes have reduced poverty levels among older Americans. But elderly women—particularly those who are divorced or widowed—are not able to access benefits of public policies. Morris (2007) also mentioned that women face greater economic challenges in retirement than men as they live longer and have lower lifetime earnings than men; thus, they face retirement with smaller pensions and other assets than men. Najjumba-Mulindwa (2003) also agreed that the widowed and disabled women and those living alone are the ones most prone to chronic poverty. Furthermore, it is not solely lack of social security, but factors such as unemployment, chronic ill health, lack of skills, HIV/AIDS, low land productivity, political instability, low agricultural returns, and functional inability due to old age that are responsible for chronic poverty among the aged. According to Walker (1981), poverty in old age is related to the low economic and social status prior to retirement and low level of state support.

In developing countries, the older people are typically one of the poorest groups of society (Najjumba-Mulindwa 2003), and old age is associated with a higher incidence of financial vulnerability and social exclusion (Barrientos et al. 2003). According to a Planning Commission of India report (GoI 2011), as people get older, they need more intensive and long-term care, which in turn may increase financial stress on the family; consequently, India's older people constitute a major proportion of those enduring chronic poverty. Pal and Palacios (2011) examined whether the households with elderly members have higher rates of poverty than that with non-elderly members. They have found that households with elderly members do not necessarily have higher poverty rates than the non-elderly household. However, the increased dependency rate of poor households could have some important effects on the consumption of the rest of the members of the household. This could be offset, to some extent, if the older family members contribute to the overall income of the household (Pal and Palacios 2006).

Srivastava and Mohanty (2012) have estimated poverty in three different ways:

(i) by applying an official cut-off point of the poverty line to household consumption expenditure
(ii) by adjusting consumption expenditure to household size
(iii) by adjusting consumption expenditure to household composition.

They reported that 18 million elderly people in India are living below the poverty line. By using the last two above-mentioned methods, they have observed that there are no significant differences in the incidence of poverty among elderly and non-elderly households in India. The study also found that poverty among the elderly living with non-elderly members is lower compared to the elderly living with elderly members. According to them, one of the important predictors of poverty among the elderly is the level of education of the aged. Srivastava and Mohanty (2012) depict that poverty increases more with an increase in the number of elderly in a household in a rural area compared to the urban area may be because of greater economic independence of the elderly in urban areas.

4.5.1 *Effects of Contribution of Elderly on Household Poverty*

The aged are often treated as an economic burden in the literature on gerontology, particularly as their healthcare costs are often substantial (Nyce and Schieber 2005), and may push the household below the poverty line (Jinkook and Drystan 2011). In this section, we want to examine whether the contribution of the elderly significantly helps to reduce the household poverty rates in rural and urban India and also in different states across the place of residence in recent years compared to the period just after globalization. For that purpose, first of all, we have calculated the HCR of the household considering MPCE in each round (NSS 55th and NSS 68th

Table 4.4 Head count ratio of rural and urban India in the 55th and 68th rounds of NSS

55th				68th			
HCR	HCR'	% change	t-statistics	HCR	HCR'	% change	t-statistics
Rural							
24.68	33.14	8.45	−35.24	19.08	27.99	8.91	−36.29
Urban							
20.32	25.31	4.99	−18.58	16.16	22.36	6.20	−22.77

Source: Calculated from NSS 55th and 68th rounds

round) by using Planning Commission's estimate of the poverty line. In the next step, the HCR (which is indicated by HCR') has once again been calculated in both the rounds considering the variable MPCE' as explained in the methodology section by using Planning Commission's estimate of the poverty line. The difference between two HCRs provides an estimate of the extent to which the aged workers are able to pull up households above the poverty line. We have estimated 't'-statistics to check whether the difference between HCR and HCR' is statistically significant.

The estimate of household poverty (HCR, HCR') in the 55th and 68th rounds (Table 4.4) reveals that rural poverty is higher than urban poverty, and over the study period, household poverty has reduced more in rural areas than in urban India (comparing HCR in the 55th and 68th rounds). This may be because of the creation of job opportunities in rural India particularly with the inception of NREGA, and success of public distribution system. It is also possible that the decline in rural poverty had a positive impact on urban well-being. Rural prosperity may have reduced internal rural to urban migration, thereby helping to reduce poverty in urban India. Datt and Ravallion (2011), however, argue that, in the post-reform period, there is much stronger evidence of a feedback effect from urban economic growth to rural poverty reduction. They have also claimed that post-reform rural growth has been less poverty reducing in a rural area but the rising overall living standards in India's urban areas in the post-reform period appear to have significant distributional effects favoring the country's rural poor.

Table 4.4 also illustrates that the contribution of elderly significantly helps to improve the household poverty rate in rural India and in urban India in the 55th and 68th rounds of NSS. However, household's economic status depends on the financial contribution of elderly more in the 68th round compared to the 55th round. Without the net contribution of the elderly, the proportion of household under the poverty line was found to increase more in 2011–2012 compared to 1999–1900. While this was found in both rural and urban India, the increase was higher in rural India, particularly in the 68th round. This may be because of the fact that in rural India, the income may have increased more among the elderly who belongs to just below the poverty line; on the other hand, in urban areas, income has increased mainly for the above poverty line elderly or who belong to far below the poverty line. Another reason may be that the contribution of rural elderly is much higher than the urban elderly.

4.6 State-Wise Analysis of the Effects of Contribution on Household Poverty

In the next part, we have analysed whether the contribution of the elderly population to their families has had any impact on the incidence of poverty in different states of rural and urban India over the study period.

4.6.1 State-Wise Analysis of the Effects of Contribution of Elderly Workers on Household Poverty in the 55th Round

The HCR for the entire sample and for the sub-sample excluding aged members (and their net contribution) for different states in rural and urban areas of India has been reported in Appendix Table 4.8. Choropleth maps are used to identify regional patterns in *changes* in HCR across rural and urban areas for the two rounds.

The choropleth map depicting difference between HCR and HCR′ in rural India in the 55th round shows that the poverty of the household increases in all the states without the contribution of elderly (Table 4.8). Figure 4.13 illustrates the increase in household poverty in terms of increase in HCR in different states of rural India in 1999–2000. The increase in household poverty is very high in the rural areas of states such as Uttar Pradesh, Himachal Pradesh, Punjab, Rajasthan, Kerala, and Maharashtra in 1999–2000. The lowest increase in HCR has been found in Karnataka, Orissa, Bihar, Jharkhand, West Bengal, Haryana, and the North-Eastern states.

Excluding contribution of the elderly members increases household poverty levels significantly in urban areas of almost all states in 1999–2000. In urban India, excluding contribution of elderly members increased poverty levels most in Kerala, Tamil Nadu, Haryana, Punjab, Uttar Pradesh, Uttarakhand, Himachal Pradesh, West Bengal, and Gujarat. On the other hand, the increase in HCR has been relatively low in Maharashtra, Madhya Pradesh, Chhattisgarh, Orissa, Andhra Pradesh, Jammu and Kashmir, and the North-Eastern states (Fig. 4.14).

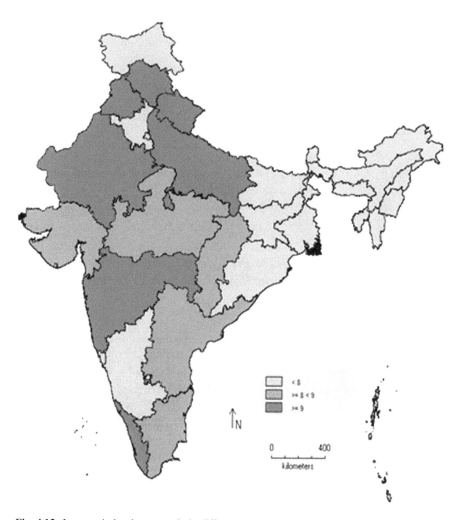

Fig. 4.13 Increase in head count ratio in different states after excluding income of aged in rural India in the 55th round (percent). (Source: NSS 55th round)

4.6.2 State-Wise Analysis of the Effects of Contribution of Elderly Workers on Household Poverty in the 68th Round

State-wise poverty levels for the 68th round, measured in terms of HCR, are given for rural and urban areas in Appendix Table 4.9. Choropleth maps, showing state-wise change in HCR (after dropping contribution of aged members), are given in Figs. 4.15 and 4.16.

4.6 State-Wise Analysis of the Effects of Contribution on Household Poverty

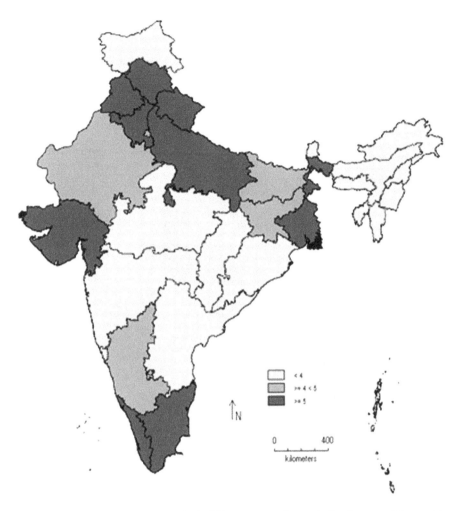

Fig. 4.14 Increase in head count ratio in different states after excluding income of the aged in urban India in the 55th round (percentage). (Source: NSS 55th round)

Comparing the HCR and HCR′ in the 68th round in rural India, we observe that household poverty increases significantly in almost all the states, if we exclude the net contribution of the elderly workers (Table 4.9). The highest increase is observed in Himachal Pradesh, Punjab, Rajasthan, Uttar Pradesh, Andhra Pradesh, Tamil Nadu, and Kerala. The increase is relatively low in states such as Uttarakhand, Madhya Pradesh, Chhattisgarh, Jharkhand, West Bengal, and the North-East regions of India.

In urban India, in 2011–2012, the increase in poverty is relatively high in Kerala, Tamil Nadu, Maharashtra, Gujarat, Rajasthan, Punjab, and West Bengal, while the decrease in HCR is relatively low in Himachal Pradesh, Uttarakhand, Madhya Pradesh, Chhattisgarh, Orissa, Jharkhand, and parts of the North-East region.

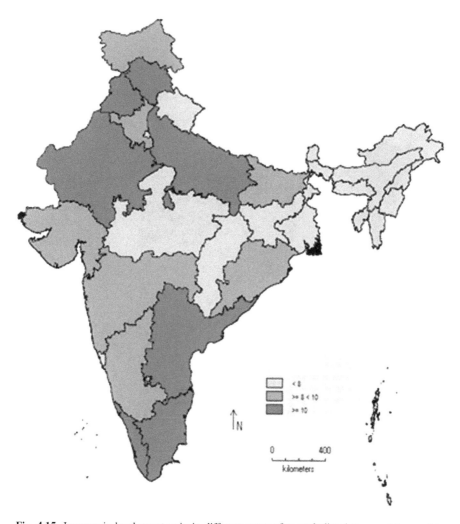

Fig. 4.15 Increase in head count ratio in different states after excluding income of the aged in rural India in the 68th round (percentage). (Source: NSS 68th round)

4.7 Change in Household Poverty

4.7.1 Change in Household Poverty in the 55th Round: Link with Proportion of Elderly and Log of State Domestic Product

Figure 4.17 indicates a positive relation between the proportion of elderly and percentage change of household poverty in both rural and urban India in the 55th round of NSS. This implies that higher the proportion of elderly in the states, higher the impact of the contribution of elderly on the change in household poverty. A positive relationship is also observed between the changes in HCR and log of net state

4.7 Change in Household Poverty

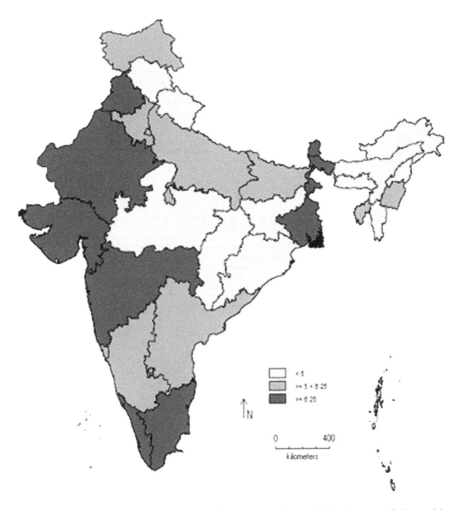

Fig. 4.16 Increase in head count ratio in different states after excluding income of the aged in urban India in the 68th round (percentage). (Source: NSS 68th round)

domestic product in rural and urban India in the 55th round of NSS (Fig. 4.18). However, this positive trend is clearer in rural India, compared to that in urban India.

4.7.2 Change of Household Poverty in the 68th Round: Link with Proportion of Elderly and Log of State Domestic Product

In the 68th round, on the other hand, although a positive relationship between the change in HCR and the proportion of elderly may be seen in urban areas, there is no such clear relationship between the two in rural India. With respect to net state

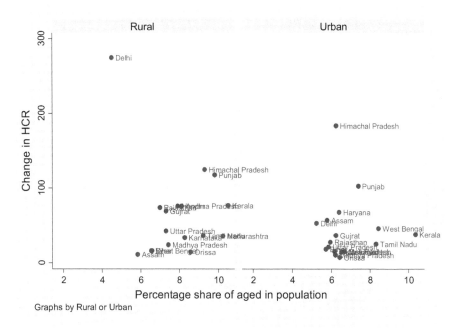

Fig. 4.17 Change in poverty in different states across the proportion of elderly in rural and urban India in the 55th round of NSS (percentage). (Source: NSS 55th round)

domestic product, however, a positive relation is again observed in rural and urban India (Figs. 4.19 and 4.20).

4.8 State-Wise Analysis of the Effects of Financial Contribution of Elderly Workers on the Intensity of Household Poverty

As we have already mentioned, a major limitation of HCR as a measure of poverty is that it captures only the number of poor; it does not reflect the intensity of poverty. For this purpose, the FGT measure is a more appropriate measure. We have calculated two measures of FGT, FGT2, and FGT2′. While FGT2 is the poverty measure for the complete NSSO sample household, FGT2′ has been calculated after excluding the net contribution of the aged, and for only the non-aged sub-sample.

4.8 State-Wise Analysis of the Effects of Financial Contribution of Elderly Workers... 79

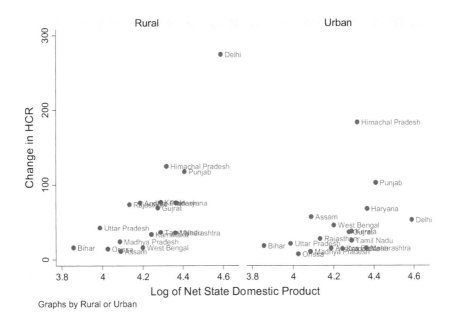

Fig. 4.18 Change in poverty in different states across the net state domestic product in rural and urban India in 55th round of NSS (percentage). (Source: NSS 55th round)

4.8.1 Analysis of FGT Index in the 55th Round

Without the net contribution of the aged, the intensity of poverty increases in all states in rural India in the 55th round of NSS (Appendix Table 4.10). According to FGT2 index, in rural India (Fig. 4.21) in the 55th round, the contribution of elderly helps to reduce the intensity of household poverty remarkably in Kerala, Karnataka, Andhra Pradesh, Tamil Nadu, Uttar Pradesh, Himachal Pradesh, Punjab, and Uttarakhand. The effects of elderly contribution on the intensity of poverty are much less in rural Rajasthan, Madhya Pradesh, Chhattisgarh, Orissa, West Bengal, and North-Eastern states.

In urban India, the impact of excluding elderly members on the intensity of poverty is high in Assam, Haryana, West Bengal, Andhra Pradesh, Kerala, and Tamil Nadu. In states such as Rajasthan, Madhya Pradesh, Chhattisgarh, Karnataka, and Jammu and Kashmir, as well as North-Eastern parts, the decline in intensity of poverty after dropping the aged respondents is relatively low (Fig. 4.22).

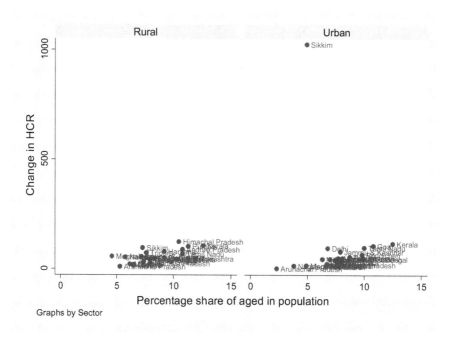

Fig. 4.19 Change in poverty in different states across the proportion of elderly in rural and urban India in the 68th round of NSS (percentage). (Source: NSS 68th round)

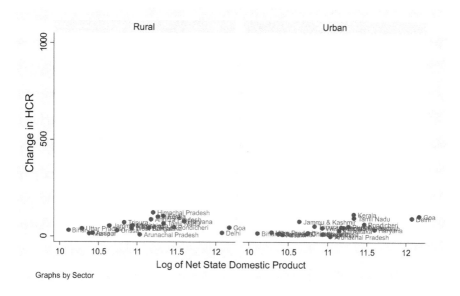

Fig. 4.20 Change in poverty in different states across the net state domestic product in rural and urban India in 68th round of NSS (percentage). (Source: NSS 68th round)

4.8 State-Wise Analysis of the Effects of Financial Contribution of Elderly Workers... 81

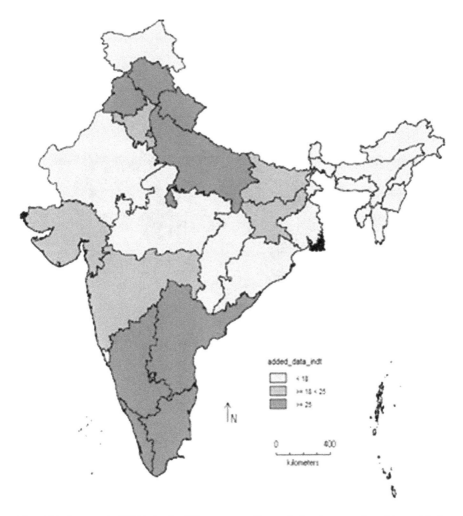

Fig. 4.21 Increase in FGT index in different states after excluding income of aged in rural India in the 55th round (percentage). (Source: NSS 55th round)

4.8.2 Analysis of FGT Index in the 68th Round

The choropleth map showing the change in the FGT index in rural areas for the 68th round (Fig. 4.23) reveals that the increase in FGT after dropping the aged persons is highest in Jammu and Kashmir, Haryana, Himachal Pradesh, Jharkhand, Orissa, Andhra Pradesh, Kerala, and Assam. The increase in the intensity of rural poverty without elderly contribution is relatively low in the states such as Maharashtra, Rajasthan, Bihar, and West Bengal.

Figure 4.24 illustrates the increase in FGT index across states in urban India for 2011–2012 after dropping the aged sub-sample. In the 68th round, the increase is

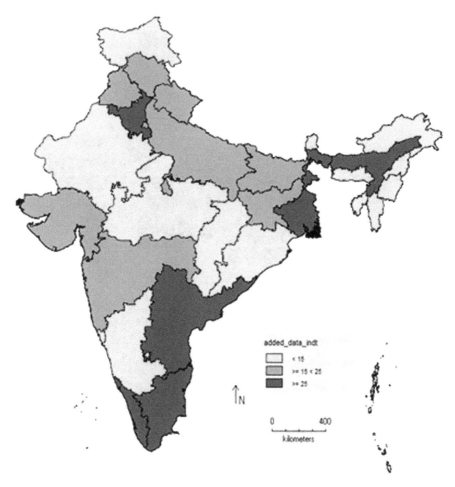

Fig. 4.22 Increase in FGT index in different states after excluding income of the aged in urban India in the 55th round (percentage). (Source: NSS 55th round)

relatively high in Jammu and Kashmir, Haryana, Himachal Pradesh, Jharkhand, Orissa, Andhra Pradesh, Kerala, and Assam. In contrast, the increase in FGT is low in the states of Rajasthan, Gujarat, Himachal Pradesh, Uttar Pradesh, Madhya Pradesh, Chhattisgarh, Bihar, and West Bengal (Table 4.11).

4.8 State-Wise Analysis of the Effects of Financial Contribution of Elderly Workers... 83

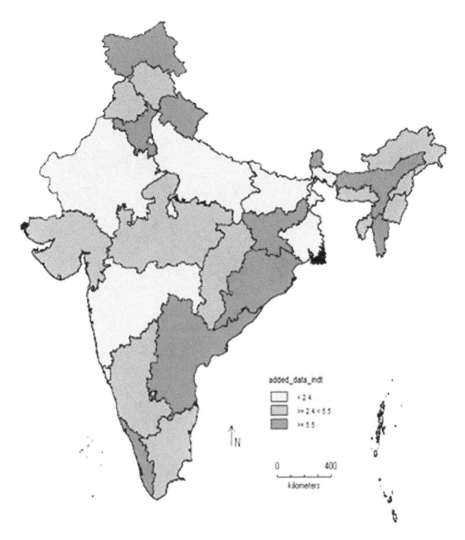

Fig. 4.23 Increase in FGT index in different states after excluding income of the aged in rural India in the 68th round (percentage). (Source: NSS 68th round)

4.8.3 FGT Index at the All-India Level

The two FGT indices, FGT2 and FGT2′, at the all-India level for the two rounds by place of residence is given in Fig. 4.25. It can be seen that excluding the aged leads to a sharp increase in the intensity of poverty. It may be seen that, in all cases, FGT2′ is higher than FGT2; furthermore, the difference between the two measures is higher in the 68th round in both rural and urban areas. The rural–urban difference in change of poverty levels, however, is marginal.

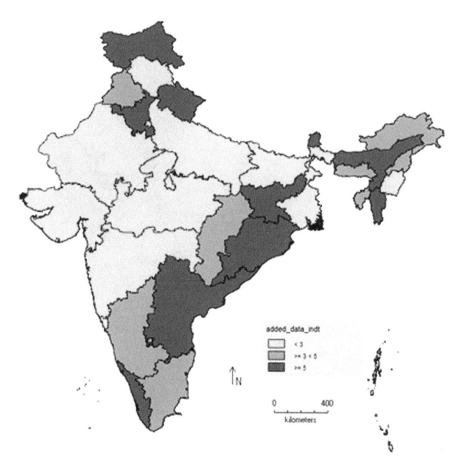

Fig. 4.24 Increase in FGT index in different states after excluding income of aged in urban India in the 68th round (percentage). (Source: NSS 68th round)

4.9 Summing Up

Studies on the aged in India and in any other countries have generally tended to portray the aged as a vulnerable section of the community requiring the support of the state and the young. This approach has often resulted in the aged being viewed as a burden on society.

In this chapter, we have examined the financial contribution of the elderly people to household's economic status, considering NSS 55th and 68th rounds. Over the study period, we have seen that gross contribution of the elderly has increased at the

4.9 Summing Up

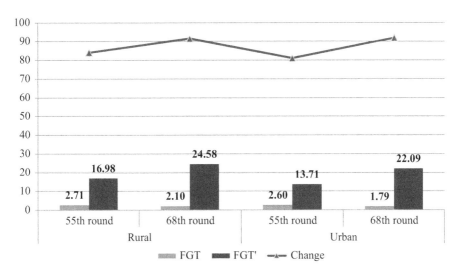

Fig. 4.25 FGT index, including and excluding contribution of the aged, by place of residence and rounds

all-India level, and also in rural and urban India. However, we have also found that the net contribution is negative in both rural and urban India, and in both rounds.

Econometric analysis reveals that the household expenditure level does not have any significant relation with net contribution of the elderly in rural and urban India in both the rounds. Net contribution is found to be higher in larger families, and in families with a smaller number of working members. In both the rounds in the rural and urban areas, weaker social group (STs and OBCs households) are more dependent on the aged compared to that of Hindu Upper Caste families. In rural areas, net contribution is higher in families engaged in agriculture. In urban areas, families with a higher proportion of self-employed are most dependent on the aged. State-level unemployment rates are negatively related to the net contribution of the elderly. Per capita NSDP shows insignificant relation in the 55th round but a negative significant relation with respect to net contribution in the 68th round.

Analysis of the impact of the contribution of elderly on household poverty shows that without the contribution of elderly workers, incidence of household poverty increases both in rural and in urban India in the 55th and 68th rounds. This is confirmed by the analysis based on choropleth maps. Apart from the incidence of poverty, we have examined the intensity of poverty and have found that, without contribution of the elderly to household expenditure, intensity and incidence of poverty increases in both the rounds.

So, summing up, we have found that aged workers play a critical role in reducing the level and especially the intensity of poverty in rural and urban India in both the rounds. So, the aged should be valued not only for their past role but also for their current economic contributions.

Appendix

Table 4.5 Net contribution of elderly to household expenditure of different states in India in the 55th and 68th rounds of NSS

State	Rural India		Urban India	
	NSS 55th	NSS 68th	NSS 55th	NSS 68th
Andaman and Nicobar Islands	−7.29	−11.7	−16.48	−25.6
Andhra Pradesh	−13.91	−10.9	−15.05	−17.8
Arunachal Pradesh	−1.41	−2.5	−19.52	−2.7
Assam	−6.03	−7.3	−10.97	−20.8
Bihar	−1.56	−3.8	−7.03	−7.2
Chandigarh	1.54	0.4	−13.93	−16.3
Chhattisgarh	-	−2.1	-	−12.9
Dadra and Nagar Haveli	−19.60	−5.6	−14.35	−18.8
Daman and Diu	−22.65	−3.0	−3.58	−12.1
Delhi	−10.56	13.7	−10.14	−12.7
Goa	−17.05	−23.8	−12.71	−24.8
Gujarat	−9.46	−8.8	−15.36	−17.5
Haryana	−12.34	−6.5	−6.20	−16.1
Himachal Pradesh	1.84	−6.8	−12.27	−18.8
Jammu and Kashmir	1.22	−3.3	−12.96	−12.6
Jharkhand	-	2.3	-	−15.2
Karnataka	−10.07	−9.1	−16.94	−10.9
Kerala	−9.89	−12.3	−11.82	−14.2
Lakshadweep	−2.69	−23.8	-	4.0
Madhya Pradesh	−4.53	−2.3	−10.15	−15.1
Maharashtra	−10.81	−12.4	−13.30	−18.1
Manipur	1.69	−0.8	6.29	−3.9
Meghalaya	−4.66	2.5	−19.70	−8.8
Mizoram	−11.19	−12.7	−9.55	−6.8
Nagaland	4.77	3.8	1.11	−23.3
Orissa	−9.70	−9.7	−15.09	−13.2
Pondicherry	−13.40	−18.1	−19.87	−21.7
Punjab	−7.01	−7.1	−11.58	−11.6
Rajasthan	−4.41	−1.2	−14.88	−11.9
Sikkim	−1.63	−5.9	−12.76	−15.5
Tamil Nadu	−13.64	−15.6	−14.71	−15.7
Tripura	−4.98	−6.0	−8.10	−19.1
Uttar Pradesh	2.25	3.9	−9.82	−7.2
Uttarakhand	-	−12.7	-	−14.7
West Bengal	−7.37	−12.3	−14.80	−17.6

Source: Calculated on the basis of NSS 55th and 68th rounds data

Appendix

Table 4.6 Estimated 't' value for the contribution of elderly to household expenditure of different states in India in the 55th and 68th rounds of NSS

State	Rural India				Urban India			
	NSS 55th		NSS 68th		NSS 55th		NSS 68th	
	T	Prob.	t	Prob.	t	Prob.	t	Prob.
Andaman and Nicobar Islands	−1.90	0.06	−0.38	0.71	0.06	0.95	−2.99	0.01
Andhra Pradesh	−0.68	0.49	2.29	0.02	−3.94	0.00	−2.40	0.02
Arunachal Pradesh	2.30	0.02	−0.30	0.76	−4.53	0.00	0.12	0.91
Assam	0.84	0.40	−0.56	0.57	0.74	0.46	−3.47	0.00
Bihar	7.68	0.00	7.45	0.00	1.49	0.14	2.33	0.02
Chandigarh	−1.75	0.10	−0.48	0.64	0.01	0.99	−3.54	0.01
Chhattisgarh			−1.10	0.27			−1.98	0.05
Dadra and Nagar Haveli	−5.59	0.00	−0.08	0.94	−0.98	0.37	−0.65	0.54
Daman and Diu	−3.34	0.00	−1.68	0.14	−0.85	0.40	−1.34	0.20
Delhi	−0.39	0.70	−0.58	0.58	−0.91	0.36	−1.55	0.12
Goa	−1.78	0.08	−2.47	0.02	−0.75	0.46	−3.61	0.00
Gujarat	−2.97	0.00	−2.71	0.01	−4.71	0.00	−3.81	0.00
Haryana	−2.49	0.01	0.62	0.53	−0.12	0.91	−4.95	0.00
Himachal Pradesh	2.67	0.01	−1.41	0.16	−0.72	0.48	−2.20	0.03
Jammu and Kashmir	1.77	0.08	2.65	0.01	−2.39	0.02	−1.77	0.08
Jharkhand			5.09	0.00			0.51	0.61
Karnataka	−1.88	0.06	−1.25	0.21	−5.69	0.00	−0.91	0.36
Kerala	1.97	0.05	2.58	0.01	−2.34	0.02	1.89	0.06
Lakshadweep	0.11	0.91	−1.85	0.08			0.88	0.39
Madhya Pradesh	3.16	0.00	2.88	0.00	−1.96	0.05	−2.19	0.03
Maharashtra	0.21	0.83	−2.07	0.04	−1.48	0.14	−6.72	0.00
Manipur	3.19	0.00	1.73	0.09	2.76	0.01	2.33	0.02
Meghalaya	−3.08	0.00	−2.50	0.01	−0.42	0.68	0.27	0.79
Mizoram	−2.76	0.01	−0.04	0.97	0.21	0.84	−1.13	0.26
Nagaland	2.19	0.03	0.03	0.98	1.68	0.12	−1.59	0.13
Orissa	1.90	0.06	2.26	0.02	−2.55	0.01	0.29	0.77
Pondicherry	0.84	0.41	−0.98	0.34	−1.84	0.07	−1.63	0.11
Punjab	1.38	0.17	1.40	0.16	−4.75	0.00	−0.04	0.97
Rajasthan	−1.92	0.06	2.73	0.01	−2.50	0.01	−0.63	0.53
Sikkim	0.08	0.94	−2.40	0.02	−0.61	0.54	−1.87	0.09
Tamil Nadu	2.35	0.02	3.82	0.00	−1.06	0.29	0.86	0.39
Tripura	2.76	0.01	2.69	0.01	0.18	0.86	0.58	0.57
Uttar Pradesh	12.27	0.00	11.00	0.00	−1.14	0.26	0.90	0.37
Uttaranchal			−0.23	0.82			−3.12	0.00
West Bengal	1.74	0.08	0.05	0.96	−1.98	0.05	−2.60	0.01

Source: Calculated on the basis of NSS 55th and 68th rounds data

Table 4.7 Proportion of elderly in the population

State	2001 Rural	2001 Urban	2011 Rural	2011 Urban
Andaman and Nicobar Islands	5.49	3.61	7.60	5.20
Andhra Pradesh	8.13	6.16	10.80	7.70
Arunachal Pradesh	5.28	1.73	5.30	2.30
Assam	5.86	5.79	6.50	7.50
Bihar	6.55	5.72	7.40	7.10
Chandigarh	3.23	5.19	3.80	6.40
Chhattisgarh	–	-	8.20	6.80
Dadra and Nagar Haveli	4.29	3.02	4.70	3.30
Daman and Diu	3.74	7.45	5.90	4.30
Delhi	4.53	5.24	6.50	6.80
Goa	9.08	7.58	11.90	10.80
Gujarat	7.31	6.23	8.30	7.40
Haryana	7.93	6.41	9.20	7.70
HP	9.31	6.25	10.50	7.80
Jammu and Kashmir	6.76	6.36	7.20	7.90
Jharkhand	–	–	7.30	6.60
Karnataka	8.28	6.53	10.40	8.00
Kerala	10.52	10.36	12.60	12.50
Lakshadweep	5.65	6.77	7.80	8.30
Maharashtra	10.24	6.68	11.30	8.10
Manipur	6.41	7.54	6.50	8.20
Meghalaya	4.60	4.41	4.60	4.90
Mizoram	5.74	5.29	6.20	6.30
Madhya Pradesh	7.43	6.20	8.00	7.60
Nagaland	4.94	2.63	5.80	3.80
Orissa	8.58	6.43	9.80	7.80
Pondicherry	8.25	8.35	9.20	9.80
Punjab	9.82	7.39	11.30	8.70
Rajasthan	6.99	5.95	7.60	7.00
Sikkim	5.55	3.91	7.30	4.90
Tamil Nadu	9.23	8.31	10.80	10.00
Tripura	7.18	7.70	7.60	8.70
Uttar Pradesh	7.31	5.86	8.00	6.70
Uttaranchal	–	–	9.6	7.4
West Bengal	6.59	8.44	7.9	9.8

Source: Calculated on the basis of Census data in 2001 and 2011 Note: Census Tables C9 and C14 have been used for 1999–2000 and 2011–2012, respectively

Appendix

Table 4.8 Head count ratio in rural and urban India in the 55th round of NSS

State	HCR Rural	HCR Urban	HCR1 Rural	HCR1 Urban
Andhra Pradesh	11.38	25.94	20.03	29.9
Assam	41.098	5.75	45.75	9.04
Bihar	43.85	24.39	51.03	28.9
Gujarat	12.97	16.02	21.95	21.91
Haryana	10.34	9.34	18.197	15.66
Himachal Pradesh	11.62	3.14	26.12	8.9
Karnataka	20.56	28.42	27.52	32.55
Kerala	12.52	19.83	22.13	27.29
Madhya Pradesh	36.14	36.796	44.9	40.67
Maharashtra	25.38	25.84	34.46	29.77
Orissa	45.24	35.46	51.82	38.02
Punjab	9.06	5.95	19.72	12.06
Rajasthan	13.18	17.92	22.93	22.898
Tamil Nadu	22.62	24.61	30.93	30.81
Uttar Pradesh	28.65	28.09	40.89	34.02
West Bengal	32.7	12.81	38.02	18.69
Delhi	2.12	8.73	7.94	13.36

Source: Calculated on the basis of NSS 55th round data

Table 4.9 Head count ratio in rural and urban India in the 68th round of NSS

State	HCR Rural	HCR Urban	HCR1 Rural	HCR1 Urban
Jammu and Kashmir	15.32	7.62	23.89	13.61
Himachal Pradesh	10.68	9.92	23.96	14.62
Punjab	10.76	14.37	21.71	22.03
Haryana	11.94	13.73	21.35	18.88
Delhi	8.20	6.29	9.84	12.26
Rajasthan	19.91	16.43	30.07	23.8
Uttar Pradesh	27.05	26.92	38.05	33.01
Bihar	27.46	30.2	36.13	35.32
Sikkim	7.24	1.0	14.14	11.25
Arunachal Pradesh	26.71	20.89	29.43	21.22
Nagaland	11.46	16.19	17.56	18.47
Manipur	36.70	32.52	42.01	37.58
Mizoram	20.16	8.04	24.38	11.61
Tripura	10.75	10.48	18.60	16.18
Meghalaya	6.78	16.09	10.63	18.56
Assam	25.50	24.399	29.95	27.04
West Bengal	18.47	16.55	26.26	24.12

(continued)

Table 4.9 (continued)

State	HCR Rural	HCR Urban	HCR1 Rural	HCR1 Urban
Jharkhand	–	29.65	-	33.74
Odisha	29.19	26.16	38.37	30.43
Chhattisgarh	–	20.68	–	24.90
Madhya Pradesh	31.55	23	39.41	27.68
Gujarat	19.04	16.82	28.04	23.75
Maharashtra	21.28	12.82	31.15	19.18
Andhra Pradesh	11.76	11.68	22.16	17.496
Karnataka	20.31	18.46	29.10	24.41
Goa	10.63	5.56	15.63	11.46
Kerala	10.59	9.44	21.79	20.35
Tamil Nadu	16.12	9.44	27.00	18.63
Puducherry	19.53	10.94	28.91	18.08

Source: Calculated on the basis of NSS 68th round data

Table 4.10 FGT index in rural and urban India in the 55th round of NSS

State	FGT Rural	FGT Urban	FGT1 Rural	FGT1 Urban
Andhra Pradesh	0.68	1.81	28.71	38.86
Assam	2.43	0.24	9.04	49.36
Bihar	2.79	1.71	22.63	22.06
Gujarat	0.62	0.93	25.47	21.83
Haryana	0.46	0.6	19.74	41.47
Himachal Pradesh	0.75	0.13	51.88	20.64
Karnataka	1.1	2.76	30.78	10.87
Kerala	0.69	1.53	67.396	33.71
Madhya Pradesh	3.02	4.06	19.71	17.35
Maharashtra	1.54	2.53	22.96	19.53
Orissa	3.52	3.25	16.93	13.19
Punjab	0.78	0.34	51.75	19.43
Rajasthan	0.53	0.85	17.58	13.43
Tamil Nadu	1.52	2.08	27.51	30.93
Uttar Pradesh	1.61	1.87	30.097	23.61
West Bengal	1.77	0.83	15.19	39.73
Delhi	0.07	0.62	28.86	25.24

Source: Calculated on the basis of NSS 55th round data

Appendix

Table 4.11 FGT index in rural and urban India in the 68th round of NSS

State	FGT		FGT1	
	Rural	Urban	Rural	Urban
Jammu and Kashmir	0.526	1.194	21.553	17.273
Himachal Pradesh	2.563	4.133	5.189	6.467
Punjab	1.315	3.283	6.726	7.414
Uttarakhand	1.554	3.28	10.81	10.987
Haryana	1.964	3.418	9.807	9.772
Delhi	0.281	0.265	1.11	1.089
Rajasthan	0.654	0.762	1.898	1.972
Uttar Pradesh	1.175	2.82	2.806	4.418
Bihar	0.518	1.108	2.776	3.294
Sikkim	0.429	1.005	6.871	6.422
Arunachal Pradesh	0.825	1.334	6.152	6.302
Nagaland	1.463	4.82	5.647	8.604
Manipur	0.871	2.293	3.299	4.544
Mizoram	0.649	0.563	10.305	10.524
Tripura	2.595	4.479	7.22	8.549
Meghalaya	1.518	2.713	5.113	5.925
Assam	1.499	1.973	15.953	15.528
West Bengal	0.141	1.055	0.416	1.385
Jharkhand	1.105	1.621	20.413	18.464
Odisha	0.229	0.286	19.393	18.676
Chhattisgarh	1.905	4.48	7.367	8.979
Madhya Pradesh	0.452	0.757	3.331	3.636
Gujarat	1.364	2.055	4.236	4.866
Maharashtra	0.157	1.143	1.646	2.289
Andhra Pradesh	0.948	1.256	11.029	10.503
Karnataka	0.38	0.889	4.097	4.256
Goa	0.705	1.809	2.561	3.679
Kerala	1.841	4.25	12.484	13.029
Tamil Nadu	0.885	2.635	5.334	6.346
Puducherry	1.157	1.186	1.176	1.203

Source: Calculated on the basis of NSS 68th rounds data

References

Auer, P., and M. Fortuny. 2000. *Ageing of the labour force in OECD countries: Economic and social consequences*. Geneva: International Labour Office, Employment paper.

Barrientos, A., M. Gorman, and A. Heslop. 2003. Old age poverty in developing countries: Contributions and dependence in later life. *World Development* 31 (3): 555–570.

Datt, G., and M. Ravallion. 2011. Has India's economic growth become more pro-poor in the wake of economic reforms? *The World Bank Economic Review* 25 (2): 157–189.

Foster, J., J. Greer, and E. Thorbecke. 1984. A class of decomposable poverty measures. *Econometrica* 52 (3): 761–766.

James, K.S. 1994. Indian elderly: Asset or liability. *Economic & Political Weekly* XXIX (36): 2335–2339.

Jinkook, L., and P. Drystan. 2011. *Income and poverty among older Koreans, relative contributions of and relationship between public and family transfers*, RAND Labour and Population, Working paper.

Milbourne, P., and S. Doheny. 2012. Older people and poverty in rural Britain: Material hardships, cultural denials and social inclusions. *Journal of Rural Studies* 28 (4): 1–9.

Morris, J.L. 2007. Explaining the elderly feminization of poverty: An analysis of retirement benefits, health care benefits, and elder care-giving. *Notre Dame Journal of Law, Ethics and Public Policy* 21 (2): 571–607.

Najjumba-Mulindwa, I. 2003. *Chronic poverty among the elderly in Uganda: Perceptions, experiences and policy issues*, Available from: http://citeseerx.ist.psu.edu/viewdoc/download?doi=1 0.1.1.908.4886&rep=rep1&type=pd. Accessed on 08 Apr 2014.

Nyce, S.A., and S.J. Schieber. 2005. *The economic implications of aging societies: The costs of living happily ever after*. Cambridge: Cambridge University Press.

Pal, S., and R. Palacios. 2006. *Old age poverty in the Indian states: What the household data can say?* World Bank: Pension Reform Primer Working Paper Series.

———. 2011. Understanding poverty among the elderly in India: Implications for social pension policy. *Journal of Development Studies* 47 (7): 1017–1037.

Park, N.H., Y.J. Yeo, K.Y. Kim, W.S. Lim, Y.K. Song, and S.Y. Park. 2003. The status of poverty policy and development: Focus on income security. Yeongi-gun, South Korea: Korea Institute of Health and Social Affairs, Research report 2003–11.

Planning Commission. 2011. *A study of effectiveness of social welfare programmes on senior citizen in rural Rajasthan, Chhattisgarh, and Gujarat*. New Delhi: Government of India.

Selvaraj, S., and A. Karan, and S. Madheswaran. 2011. *Elderly workforce participation, wage differentials and contribution to household income*. New Delhi: UNFPA, Working paper 4.

Srivastava, A., and S.K. Mohanty. 2012. Poverty among elderly in India. *Social Indicators Research* 109 (3): 493–514.

Subaiya, L., and D. W. Bansod. 2011. *Demographics of population ageing in India*. New Delhi: UNFPA, Working Paper 1.

UNFPA. 2012. *Report on the status of elderly in selected states of India*, 2011. New Delhi: United Nations Population Fund.

Walker, A. 1981. Towards a political economy of old age. *Ageing and Society* 1 (1): 73–94.

Chapter 5
Economic Contribution of the Aged: A Regional Profile

5.1 Introduction

The all-India analysis undertaken in the previous chapter focused on the economic contribution of the aged. This chapter continues to focus on economic contribution, but it undertakes the analysis for three states of Eastern India—West Bengal, Orissa, and Jharkhand. This creates the foreground for assessing the net economic contribution of the aged in the closing sections of this chapter.

5.2 Workforce Participation

Workforce participation among the aged is found to be very low. Out of our sample of 481 persons, 17.9% were found to be engaged in economic activities, while a further 5.4% were unemployed (not working, but actively seeking work). Breaking up the sample by gender, we found that 26.5% of aged males and 9.7% of aged females were working. From Fig. 5.1, we see that workforce participation is relatively higher among males, and in Ranchi.

In Table 5.1, we present workforce participation across socio-demographic correlates. We find that workforce participation is high among the middle-income and affluent households, those engaged in the informal sector or as casual workers, among OBC households, and among non-Hindu households. Workforce participation was also observed to be relatively high among respondents aged 66–70 years.

© Springer Nature Singapore Pte Ltd. 2020
Z. Husain, *Active Ageing and Labour Market Engagement*,
https://doi.org/10.1007/978-981-15-0583-6_5

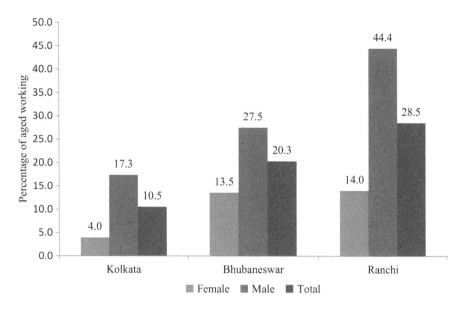

Fig. 5.1 Workforce participation by gender across cities. (Source: Estimated from primary survey data)

5.3 Nature of Employment

The majority of aged workforce are engaged as manual workers, followed by the residual category 'Others'; the proportion engaged in retail trade is also high (Fig. 5.2). In case of female respondents, the majority is engaged in manual work, followed by retail trade.

The occupational distribution of respondents varies across place of residence (Table 5.2). In Kolkata, for instance, female workers are distributed equally between service, business, and manual work, while in Bhubaneswar retail trade and business are the main occupations. In Ranchi, the majority of female workers are engaged as manual workers. Considering male workers, we find that a majority is engaged in business (Kolkata), retail trade and manual work (Bhubaneswar), and retail trade and business (Ranchi).

Most of the workers are engaged as regular workers (81%), followed by part-time workers (14%). This holds for both male and female workers.

5.4 Earnings and Gross Contribution to Household

The monthly earnings of workers was found to be ₹1517 (female) and ₹4497 (male). City-wise variation was quite substantial, with earnings of aged workers being highest in Ranchi (₹3810 and ₹12,277 for female and male workers, respectively). The majority of earning respondents do not retain this income, but they contribute their

5.4 Earnings and Gross Contribution to Household

Table 5.1 Workforce participation by socio-demographic correlates

Socio-demographic correlates	Categories	Gender of respondent		
		Female	Male	Total
Monthly expenditure group	Under 23,550	10.9	13.5	11.9
	23,551–49,950	5.4	25.4	17.7
	49,951–73,525	10.9	31.0	19.6
	73,526–104,500	10.4	18.2	14.1
	Above 104,500	9.3	40.4	26.3
Age	60–65 years	9.2	27.7	18.5
	66–70 years	12.5	32.6	22.3
	71–80 years	5.6	15.7	10.5
	81 years and above	15.4	33.3	22.7
Education	Illiterate	14.9	31.0	18.7
	Below middle	12.5	30.6	22.1
	Middle	13.0	32.4	24.6
	Secondary	3.3	32.4	18.8
	HS	5.9	27.8	20.8
	Others	2.0	15.4	9.5
Religion	Hindu	9.5	25.6	17.8
	Others	10.9	36.8	18.5
Social group	General	7.7	23.4	16.0
	SC/ST	5.6	25.0	12.8
	OBC	19.6	38.6	28.4
Household type	Self-employed	12.9	45.2	31.5
	Regular wage earner	2.7	30.4	13.4
	Casual labour	30.8	33.3	32.4
	Others	10.8	17.2	13.9

Source: Estimated from primary survey data

earnings to the family kitty. This holds particularly for female workers in all cities. In case of male workers, 7% (Kolkata) and 68% (Ranchi) keep their earnings with themselves.

These earnings comprise the gross contribution of the aged to their households. Unfortunately, the mean of this contribution is not very high as the majority of respondents do not work and earnings are low. Thus, we find that the mean gross contribution is only 4.8% of the household monthly expenditure. Predictably, the gross contribution of male respondents is higher than that of female respondents (5.7% compared to 4.4%). If we consider only households where the aged respondents are engaged in economic activities, the corresponding percentages are 6.5 and 5.1, respectively.

In Table 5.3, we present the mean gross contribution of different categories formed on the basis of socio-demographic correlates. Elderly persons contribute more in low-income households, households engaged in either informal sector or casual work, and in Ranchi. Respondents without education or with HS or higher levels of education, non-Hindus, those belonging to SCs/STs, and those in the age group of 60–65 or 71–80 years contribute more than other categories.

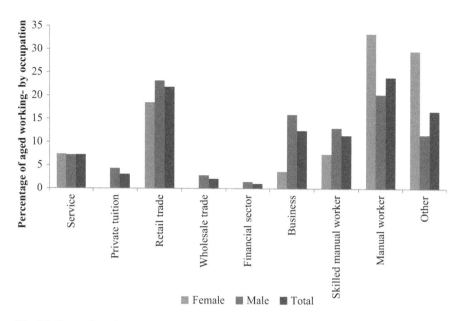

Fig. 5.2 Proportion of workers engaged in different occupation. (Source: Estimated from primary survey data)

Table 5.2 Occupational distribution of aged workers by gender and place of residence

Occupation	Kolkata		Bhubaneswar		Ranchi	
	Female	Male	Female	Male	Female	Male
Service	33.3		7.7	4.0		13.6
Private tuition		17.6				
Retail trade		17.6	30.8	36.0		18.2
Financial sector		5.9		4.0		9.1
Business	33.3	29.4	38.5	4.0		18.2
Other		17.6	7.7	16.0	30.0	13.6
Skilled manual worker		5.9	15.4	36.0	10.0	13.6
Manual worker	33.3	5.9			60.0	13.6

Source: Estimated from primary survey data

5.5 Determinants of Working, Earnings, and Gross Contribution

In the first step of our econometric analysis, we estimated a model to identify determinants of working. Since working is a binary variable (=0 if not working; = 1 if working), a logit model was estimated. Results are given in Table 5.4, with odd ratios (and not coefficients) being reported.

Male respondents, predictably, have a much higher probability of working than their female counterparts. The probability of working is significantly lower in the

5.5 Determinants of Working, Earnings, and Gross Contribution

Table 5.3 Mean of gross contribution by aged to households

Socio-demographic correlates	Categories	Female	Male	Total
Monthly household expenditure groups	Under Rs. 23,550	8.78	15.55	11.6
	23,551–49,950	0.07	4.63	4.1
	49,951–73,525	0.02	4.02	2.82
	73,526–104,500	5.81	7.15	6.59
	Above 104,500	0.46	4.57	3.66
Household type	Self-employed	5.24	7.22	6.81
	Regular wage earner	0.07	6.3	5.57
	Casual labour	14.22	2.86	6.99
	Others	0.69	4.66	3.01
Education	Illiterate	5.82	5.08	5.55
	Primary or below	0.05	9.2	6.15
	Middle	0.02	2.82	2.35
	Secondary	0.04	4.12	3.78
	Higher secondary	0.1	5.91	5.43
	Others	0.03	6.67	6.11
Age groups	60–65 years	7.81	7.32	7.44
	66–70 years	0.88	3.16	2.37
	71–80 years	0.09	7.53	5.67
	81 years and above	0.08	0.04	0.05
Religion	Others	14.09	26.46	20.27
	Hindu	0.45	3.41	2.63
Caste	SC and ST	0.08	18.89	14.55
	OBC	7.93	7.62	7.74
	General	0.51	1.08	0.94
Place of residence	Kolkata	0.09	0.16	0.15
	Bhubaneswar	0.05	0.04	0.05
	Ranchi	9.26	15.72	13.70

Source: Estimated from primary survey data

71–80 years age group but is same for other age groups. Economic status of the household (captured by monthly household expenditure) does not determine probability of working; nor does education of respondent. Only those with above school education are found to be less likely to be working. SC/ST respondents are less likely to work at a 10% level, but probability of working is same across religion, and between General castes and OBCs. Household size is not an important determinant of working. City-wise variations are observed, with respondents from Ranchi having a higher probability of working.

We next turn to a model of earnings. Given that there is non-random truncation of the sample, with a group not working and so not earning either, use of a Heckman model is recommended. Household size is used as the instrument. Results of both sections of the model (determinants of working, determinants of earning) are reported in Table 5.5.

Table 5.4 Results of logit model for determinants of working

Work	Odd ratios	z	P > z
Female (ref)	1.00		
Male	2.20	5.07	0.00
60–65 years (ref)	1.00	–	–
66–70 years	1.12	0.58	0.56
71–80 years	0.63	−2.25	0.02
81 years and above	1.17	0.64	0.52
Up to primary (ref)	1.00	–	–
Middle	1.16	0.67	0.50
Secondary	0.88	−0.52	0.60
Higher secondary	0.90	−0.39	0.69
Above higher secondary	0.56	−2.25	0.03
Monthly household expenditure	1.00	0.05	0.96
Non-Hindu (ref)	1.00	–	–
Hindu	1.14	0.52	0.60
General (ref)	1.00		
SC and ST	0.67	−1.78	0.08
OBC	1.06	0.28	0.78
Household size	0.95	−1.39	0.17
Kolkata (ref)	1.00	–	–
Bhubaneswar	1.39	1.56	0.12
Ranchi	2.41	3.22	0.00
N	508		
Pseudo-R^2	0.14		
χ^2	66.03		0.00

As seen earlier, *male respondents have a higher probability of working; however, there is no gender gap in earnings.* Age does not have an effect on either probability of working (except that those aged 71–80 years are less likely to work) nor does it affect the level of earnings. Coefficients of education, religion, caste, and household expenditure level are found to be insignificant even at 10% level. Respondents from Ranchi are found to have a higher probability of working and also earning more than respondents from other cities.

We have estimated two models to identify determinants of gross contribution (Table 5.6). The first model is estimated using the instrument variable model as gross contribution, and monthly household expenditure may have a two-way causality. Again, the household size is used as an instrument. The Wu–Hausman test supports our hypothesis of endogeneity; the *F*-test indicates that the instrument used (household size) is relevant—strongly correlated with monthly household expenditure. In addition, we also used a Tobit model, with a lower limit of zero, to identify determinants of gross contribution. Both models are qualitatively similar, so that results are robust.

5.6 Indirect Economic Contribution of Aged

Table 5.5 Heckman model for determinants of working and earnings

Variables	Work			Earnings		
	Coef.	z	$P > z$	Coef.	z	$P > z$
Male	0.77	4.81	0.00	8003.64	1.14	0.26
60–65 years (ref)	–	–	–	–	–	–
66–70 years	0.06	0.34	0.73	−857.53	−0.35	0.73
71–80 years	−0.51	−2.38	0.02	−5515.65	−1.00	0.32
81 years and above	0.16	0.66	0.51	−710.26	−0.22	0.83
Up to primary (ref)	–	–	–	–	–	–
Middle	0.09	0.38	0.70	−5.52	0.00	1.00
Secondary	−0.16	−0.64	0.52	−3361.33	−0.94	0.35
Higher secondary	−0.06	−0.21	0.83	1047.20	0.32	0.75
Higher secondary and above	−0.57	−2.17	0.03	−5328.11	−0.95	0.34
Monthly household expenditure	0.00	0.18	0.86	0.02	1.14	0.26
Non-Hindu (ref)	–	–	–	–	–	–
Hindu	0.23	0.91	0.37	2248.35	0.55	0.58
General (ref)	–	–	–	–	–	–
SC and ST	−0.35	−1.53	0.13	−4908.42	−1.20	0.23
OBC	0.05	0.26	0.80	178.96	0.07	0.95
Household size (instrument)	−0.04	−1.27	0.20	–	–	–
Bhubaneswar	0.39	1.78	0.08	2435.06	0.59	0.56
Ranchi	0.98	3.54	0.00	17,135.52	2.02	0.04
Intercept	−1.52	−4.34	0.00	−24,652.53	−1.02	0.31
Mills λ	11,558.18	1.02	0.31			
N	503					
Pseudo R^2	0.07					
Wald χ^2	22.45		0.00			

Results indicate that monthly household expenditure is positively related with gross contribution; the coefficient is significant at 1% level. *Male respondents make a higher economic contribution to their families.* Compared to respondents with no education or those with primary education, respondents with middle level of education or with above school education contribute less to households. Aged from Hindu households contribute less than their non-Hindu counterparts; caste, however, does not have any impact.

5.6 Indirect Economic Contribution of Aged

Although aged respondents are found to supplement household income by working beyond their age of superannuation, this is not their main contribution. Elderly people are often engaged in extended System of National Accounts (SNA) activities, performing household chores. This has two effects—first, it frees other members

Table 5.6 Results of models for gross contribution

Gross contribution	IV model			Tobit model		
	Coef.	z	P > z	Coef.	z	P > z
Monthly household expenditure	0.04	6.33	0.00	0.02	7.28	0.00
Male	1004.44	2.40	0.02	1259.29	2.96	0.00
Age	205.08	1.26	0.21	262.68	1.58	0.12
Age^2	−2.07	−1.76	0.08	−2.69	−2.23	0.03
Primary or below (ref)	–	–	–	–	–	–
Middle	−1585.17	−2.33	0.02	−1285.86	−1.84	0.07
Secondary	−264.80	−0.41	0.68	−55.41	−0.08	0.93
Higher secondary	−609.75	−0.85	0.39	−19.92	−0.03	0.98
Above higher secondary	−1706.16	−3.00	0.00	−1197.36	−2.08	0.04
Non-Hindu (ref)	–	–	–	–	–	–
Hindu	−1708.12	−2.75	0.01	−1408.65	−2.21	0.03
General caste (ref)	–	–	–	–	–	–
SC and ST	664.43	1.17	0.24	778.56	1.34	0.18
OBC	344.08	0.59	0.56	961.49	1.67	0.10
Intercept	−3301.13	−0.57	0.57	−3754.75	−0.64	0.52
N	513			513		
(Pseudo) R^2	0.01			0.01		
Wald (Chi^2)	82.01			92.06		
Sigma				4539.194		
Exogeneity: Wu–Hausman F	19.02		0.00			
Weak instruments: F	101.18		0.00			

(particularly females) who can enter the labour market and supplement household income (Husain and Dutta 2015); second, it increases household disposable income by saving the expenditure on paid workers, who would otherwise have to be hired to undertake the household chores.

5.6.1 Aged and Household Chores

In Table 5.6, we examine the frequency with which the aged undertake various household tasks. A gendered division of labour is clearly manifested in Table 5.7; while the majority of females cook, wash clothes, and clean the house, males tend to perform chores outside the home like daily marketing, queuing in ration shops, undertaking financial services, and paying electricity bills.

The mean number of days that respondents perform household chores is given in Fig. 5.3. It can be seen that *women contribute more than male respondents in terms of number of days*. This is because activities like cooking, washing clothes, and cleaning the house are daily activities; in comparison, male activities such as marketing, paying utility bills, performing financial services, and queuing in ration shops are performed at intervals.

5.6 Indirect Economic Contribution of Aged

Table 5.7 Frequency of performing household tasks by aged

Activity	Frequency	Female	Male	Total
Cooking	Never	27.99	69.6	48.07
	Occasionally	19.03	16.00	17.57
	Frequently	8.58	5.60	7.14
	Regularly	42.54	3.60	23.75
	N.A.	1.87	5.20	3.47
Daily marketing	Never	56.72	30.4	44.02
	Occasionally	13.06	10.8	11.97
	Frequently	13.06	20.4	16.6
	Regularly	14.55	36.8	25.29
	N.A.	2.61	1.6	2.12
Queue in ration shop	Never	70.52	54.8	62.93
	Occasionally	8.58	12.4	10.42
	Frequently	8.58	16	12.16
	Regularly	9.7	14	11.78
	N.A.	2.61	2.8	2.7
Wash clothes	Never	29.48	61.6	44.98
	Occasionally	10.07	10.4	10.23
	Frequently	14.18	9.2	11.78
	Regularly	44.78	15.2	30.5
	N.A.	1.49	3.6	2.51
Clean house	Never	43.28	76	59.07
	Occasionally	11.94	11.2	11.58
	Frequently	12.31	5.2	8.88
	Regularly	30.97	4.4	18.15
	N.A.	1.49	3.2	2.32
Dusting furniture	Never	61.94	77.2	69.31
	Occasionally	13.06	10.4	11.78
	Frequently	12.31	5.6	9.07
	Regularly	11.19	4.8	8.11
	N.A.	1.49	2	1.74
Perform financial tasks	Never	66.04	49.2	57.92
	Occasionally	18.66	11.2	15.06
	Frequently	7.46	17.2	12.16
	Regularly	4.1	20	11.78
	N.A.	3.73	2.4	3.09
Pay electricity bills	Never	88.43	51.6	70.66
	Occasionally	2.61	7.6	5.02
	Frequently	1.12	12.4	6.56
	Regularly	3.73	26.4	14.67
	N.A.	4.1	2	3.09
Pay telephone bills	Never	86.94	63.6	75.68
	Occasionally	1.12	5.6	3.28

(continued)

Table 5.7 (continued)

Activity	Frequency	Female	Male	Total
	Frequently	0.75	6	3.28
	Regularly	0.75	14.8	7.53
	N.A.	10.45	10	10.23

Source: Estimated from primary survey data

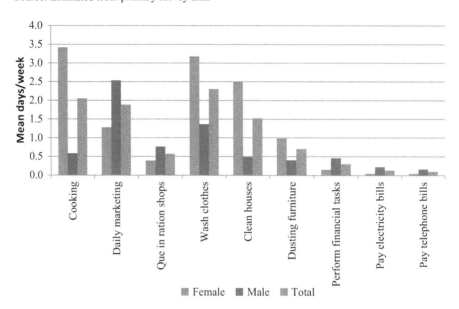

Fig. 5.3 Mean number of days in a week that a household chore is performed. (Source: Estimated from primary survey data)

Given the importance of children in the household, one would expect that the aged play a major role as informal carers to their grandchildren (Zakir Husain and Dutta 2013). Researchers have observed that reliance on formal childcare providers is costly, less reliable, and of inferior quality (compared to childcare provided by relatives); it may even adversely affect the personality development of children (Timonen et al. 2009; Griggs et al. 2010; Ruiz and Silverstein 2011). This creates an incentive to revive multigenerational ties—'relations across more than two generations' (Bengtson 2001)—and utilize the services of grandparents (Leibowitz et al. 1992; Del Boca et al. 2005; Gray 2005; Aassve et al. 2011; Zamarro 2011; Albuquerque and Passos 2012; Posadas and Vidal-Fernandez 2013). It has also been reported that the strength and direction of this relationship depend upon the sociocultural context. In Asian countries, however, the situation is more complex:

> In Asian countries elderly relatives are often conservative and persist in their stereotyped notion that women are, by nature, care providers and home builders. Work threatens this belief and is often interpreted as an attempt by modern educated women to evade their natural responsibilities to the family. The aggregation of these two conflicting forces will determine

how the presence of grandparents will affect labour market outcomes of mothers in developing countries. (Husain and Dutta 2015: 75)

Results of a primary survey of graduate women undertaken in Kolkata in 2012–2013 reported that the presence of healthy grandparents, particularly grandmothers, living close significantly increased the probability of mothers working. This finding has some important implications.

> The difficulties of balancing work and household has become a major issue in South Asian societies as the concept of household sharing of labour is yet to become popular. In particular, the responsibilities of childcare fall almost entirely on mothers. The consequent pressure on working women affects them physically and mentally, and may even lead to their withdrawal from the labour market. For instance, we found that 81 respondents (representing 17% of ever working women) had given up work after their first child because of the pressure of looking after the child; the majority (66, representing 82%) had never returned to work. The potential of grandparental supply of childcare services becomes crucial in the context of retaining women in employment in developing countries. (However) … grandparents can provide important services that contribute greatly towards family welfare. This has an important implication. When grandparents provide childcare services, they transform from "burdensome" dependents to valued members of the community. (Husain and Dutta 2015: 80-1)

Analysis of data (Table 5.8) reveals that the aged serve food to their children, take them to the doctor or nurse them (when sick), play with them, and watch TV along with them. There is a gendered division of labour with respect to medical services, with grandfathers taking the child to the doctor, but grandmothers nursing them. A large proportion of grandmothers also serve food to the children. Both grandparents, however, play with the children, or accompany them in watching TV. Interestingly, the proportion of the aged taking children to schools, or bringing them back, is quite low—only about one out of every five respondents report undertaking such chores. Nor do grandparents take the children out for extra-curricular activities or supervise their homework.

In Fig. 5.4, we have reported average number of days in a week that respondents undertake child-related activities. The mean is calculated only for those respondents who had reported performing such tasks, even on an infrequent basis. It may be seen that respondents perform child-related tasks quite frequently, except obviously for health-related services. In particular, they tend to play or watch TV with grandchildren on an almost daily basis. No gender difference is noted.

5.6.2 Imputing Economic Value to Household Chores

While studies have reported such indirect economic contributions of the aged to their families (Andrews and Hennink 1992; Hermalin et al. 1998; Randel et al. 1999; Schröder-Butterfill 2003; Barrett 2013), we have not been able to locate any study that has tried to estimate the economic value of such contributions. This is a major lacuna of studies on gerontology, given that such indirect economic contributions

Table 5.8 Frequency of performing child-related chores by aged

Activity	Frequency	Female	Male	Total
Taking children to school	Never	88.41	70.43	80.24
	Occasionally	7.25	13.91	10.28
	Frequently	1.45	6.09	3.56
	Regularly	2.9	9.57	5.93
Bringing children back from school	Never	88.89	74.56	82.33
	Occasionally	7.41	12.28	9.64
	Frequently	1.48	6.14	3.61
	Regularly	2.22	7.02	4.42
Taking children to extra-curricular activities	Never	96.75	95.28	96.07
	Occasionally	2.44	0.94	1.75
	Frequently	0	0.94	0.44
	Regularly	0.81	2.83	1.75
Supervise children's homework	Never	84.8	62.61	74.17
	Occasionally	6.4	5.22	5.83
	Frequently	6.4	4.35	5.42
	Regularly	2.4	27.83	14.58
Taking children to movies	Never	98.52	94.5	96.72
	Occasionally	0	1.83	0.82
	Frequently	1.48	0.92	1.23
	Regularly	0	2.75	1.23
Serving food to children	Never	47.86	76.11	60.47
	Occasionally	17.14	7.96	13.04
	Frequently	20.71	0.88	11.86
	Regularly	14.29	15.04	14.62
Taking children to doctor when sick	Never	80.58	58.62	70.59
	Occasionally	10.79	17.24	13.73
	Frequently	7.19	9.48	8.24
	Regularly	1.44	14.66	7.45
Nursing children when sick	Never	50	68.1	58.2
	Occasionally	15.71	13.79	14.84
	Frequently	21.43	4.31	13.67
	Regularly	12.86	13.79	13.28
Play with children	Never	45.71	48.74	47.1
	Occasionally	3.57	8.4	5.79
	Frequently	15.71	15.97	15.83
	Regularly	35	26.89	31.27
Watch TV with children	Never	52.9	44.07	48.83
	Occasionally	5.8	7.63	6.64
	Frequently	9.42	14.41	11.72
	Regularly	31.88	33.9	32.81

Source: Estimated from primary survey data

5.6 Indirect Economic Contribution of Aged

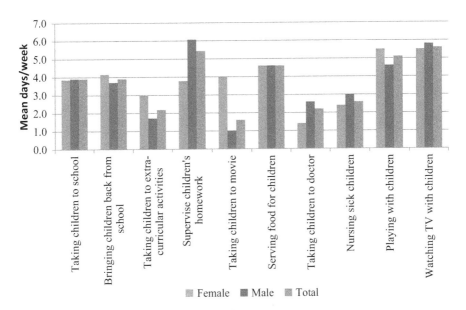

Fig. 5.4 Mean number of days in a week that a child-related chore is performed. (Source: Estimated from primary survey data)

can form a major share of total contribution of aged to their families. In this study, we have attempted to estimate the indirect economic contribution of the aged in money terms. This is not easy as there is no market for unpaid family labour, so that we can use wages or earnings to value such labour. The method employed for estimating the economic value of unpaid family labour is described below.

Our starting proposition is that the economic benefit of the aged performing a chore differs between a low-income and high-income household. In case of low-income households, when an aged member performs a household chore, this releases the female member who can seek paid work. The supplemental income is the value of non-economic contribution of the aged. On the other hand, for middle- and high-income households, the help of aged family members will allow the family to reduce dependence on paid labour. The savings from this comprises the indirect economic contribution of the aged.

Therefore, the value imputed to indirect economic contribution of the aged is the potential earnings of a female member (in case of low-income households), or the avoided expenditure on paid providers of household services. The daily value of this imputed value is multiplied by x (when $x = 30$/number of days aged provided the service). While taking the imputed value, the following points were taken into account:

(a) *Type of service provided by aged*: For instance, if the respondent paid utility bills, then the income of a driver was used for the imputation. Similarly, if the respondent cooked, the payment to a maid servant was used.

(b) *Economic status of household*: Payment to hired help will depend upon economic status of households. So we used different rates for the same service for middle- and high-income households.
(c) *City in which respondent is located*: Costs are higher in Kolkata. We used average costs in Kolkata for Kolkata respondents, and half of this for respondents in other cities.

Based on the imputation, we obtained a money value for the unpaid family labour provided by the aged. This is used to estimate net contribution of aged as follows:

$$\frac{\text{Earnings of aged} + \text{Money value of unpaid family labour} - \text{Consumption expenditure on aged}}{\text{Household expenditure}}.$$

5.6.3 Net Contribution of Aged

As we had seen previously, all aged respondents are not engaged in the labour market. In our sample, only 92 respondents are working. Their mean monthly earning is ₹3374 (Table 5.9). About two-thirds of workers are males; as expected in a male-dominated society, their earnings are higher than that of female workers. The earning gap is as high as 192% of female wages, which is substantial. If we consider the entire sample, including non-workers (and imputing their earnings as zero), mean of earnings is ₹144 and ₹1087 for female and male workers, respectively.

The mean imputed value of household services provided by the aged is ₹2037. It is interesting to note that this value is higher for male respondents. The reason lies in the gender discrimination present in the labour market; the payment for labour services provided by female workers is generally lower than that of male workers, so savings from household chores performed by female respondents is lower than that of male respondents. Mean of gross contribution of aged is about ₹2637, while mean expenditure on aged is much higher (₹11,604). As a result, mean net contribution is negative (₹−8962). A substantial gender gap is noted in all cases—*households spend more on male respondents, who make a higher gross contribution but lower net contribution.*

We have also estimated net contribution as a percentage of monthly per capita household expenditure. As the aged consume more than what they earn plus their imputed contribution to family, the mean is negative. The mean is −16.84 for the entire sample; *net contribution of male respondents (−19.40) is lower than that of females (−14.47)*. This is statistically significant ($t = 2.20$, $p = 0.3$). In Table 5.10, the percentage net contribution is given for different correlates. In all cases, it is negative. In fact, *only 17.99% of respondents make a positive net contribution to their families.*

This contradicts the perception-based results reported in an UNFPA study in India (Alam et al. 2011). In the UNFPA survey, respondents were asked whether

5.6 Indirect Economic Contribution of Aged

Table 5.9 Economic valuation of contribution of aged and expenditure on aged

Variables	Mean	Median	Standard deviation	Observations
All respondents				
Monthly earnings of aged	3374	57	6986	92
Monthly earnings of aged (all)	592	0	3202	518
Indirect contribution	2037	1072	3164	518
Total gross contribution	2637	1120	4549	518
Expenditure on aged	11,604	8199	12,056	517
Total net contribution	−8962	−6360	12,249	517
Female respondents				
Monthly earnings of aged	1429	50	3966	27
Monthly earnings of aged (all)	144	0	1311	268
Indirect contribution	1919	1280	2445	268
Total gross contribution	2063	1337	2769	268
Expenditure on aged	10,055	2614	9073	268
Total net contribution	−7992	−5567	9096	268
Male respondents				
Monthly earnings of aged	4182	60	7794	65
Monthly earnings of aged (all)	1087	0	4358	250
Indirect contribution	2165	850	3786	250
Total gross contribution	3252	976	5832	250
Expenditure on aged	13,271	9185	14,436	249
Total net contribution	−10,006	−7410	14,866	249

they contributed to their families; if so, what percent of family expenditure did they contribute. No attempt was made to verify their claims. The response was:

> 52 per cent of the elderly reported that they contribute their personal income towards the expenditure of the household. More elderly men (71%) than elderly women (36%) reported doing so … Financial contribution by the elderly does not vary very much across rural and urban areas … economic contribution of the elderly to the household expenditure varies greatly by employment status, education, living arrangement, marital status and age. The contribution of personal income decreases with age. Elderly in the age group 60-69 contribute 55 per cent of the household expenditure, those in the age group 70-79 years contribute 50 per cent, while those 80 years and above contribute 46 per cent of the household expenditure (Alam et al. 2011:51).

The report, however, reports a high level of economic dependency in a later section:

> Overall, 23 per cent of the elderly are economically independent, half are fully dependent on others, and 26 per cent are partially dependent. The proportion of elderly who are economically independent (23%), as observed in this survey, is lower than the All India rate using NSSO data for the period 2004-05 (34%). (Alam et al. 2011:51)

While the finding that aged are an economic asset to their families—contributing about half of household expenditure—is very interesting, it is contradicted by the high level of economic vulnerability reported in the same report. The inconsistency

Table 5.10 Average percentage net contribution of aged by correlates—by gender

Variables		Gender of respondent		
		Male	Female	Total
City	Kolkata	−35.25	−24.39	−29.64
	Bhubaneswar	−6.71	−4.90	−5.78
	Ranchi	−9.59	−11.72	−10.72
Age	60–65 years	−19.91	−15.06	−17.46
	66–70 years	−20.75	−13.33	−16.79
	71–80 years	−15.65	−15.90	−15.78
	81 years and above	−22.93	−11.06	−16.15
Education	Illiterate	−8.58	−4.78	−5.69
	Below middle	−12.91	−9.47	−11.19
	Middle	−15.92	−25.44	−20.10
	Secondary	−18.72	−19.67	−19.06
	HS	−31.37	−21.80	−27.19
	Others	−8.58	−4.78	−5.69
Religion	Hindu	−20.17	−14.05	−17.17
	Others	−10.16	−16.38	−14.62
Social group	General	−23.42	−19.88	−21.73
	SC/ST	−13.36	−5.80	−8.54
	OBC	−8.58	−8.36	−8.46
Household type	Self-employed	−17.06	−10.95	−14.30
	Regular wage earner	−22.37	−19.98	−20.92
	Casual labour	−13.24	−6.93	−10.98
	Others	−20.25	−13.01	−16.49

between the two results raises questions about the data quality and validity of the report. We would argue that, *in an economic sense, aged are a burden, or tax, on society.* However, we should keep in mind two points in this context. First, the substantial economic contribution made by currently aged persons in their middle ages and, in particular, *the investment made on their children; this represents a debt of the younger generation to the aged, which is repaid when the younger generation joins the labour force, replacing the superannuated aged.* Second, the *aged also makes a cultural contribution* as they pass on the culture and traditions of their society to the younger generations. There is no way to value this very important service. *If we could add these two components to net contribution, it is more than likely that we would find that aged are making a positive contribution to society.*

In metropolitan Kolkata, where cost of living is much higher than in Ranchi and Bhubaneswar, the net contribution of aged is lowest. Net contribution increases slightly with age, though there is no clear trend when we look at the age–net contribution curve by gender. The education–contribution is roughly U shaped, with net contribution being highest among respondents with no education, and respondents with above HS education. The MPCE–contribution curve displays a positive trend for male (and total) samples, with net contribution being highest for the 'richest'

5.6 Indirect Economic Contribution of Aged

Table 5.11 Results of OLS model for net contribution of aged

Variables	Net contribution (in ₹)			Net contribution (% of household expenditure)		
	Coeff.	t	P > t	Coeff.	t	P > t
Male (ref)						
Female	369.90	0.36	0.72	4.90	2.27	0.02
61–65 years (ref)						
66–70 years	−1156.92	−0.90	0.37	−0.34	−0.13	0.90
71–80 years	−266.89	−0.21	0.83	0.58	0.22	0.83
81 years and above	−3508.12	−1.98	0.05	−3.54	−0.96	0.34
Household size	182.56	0.84	0.40	1.55	3.44	0.00
Log of MPCE	−5131.74	−8.04	0.00	1.09	0.82	0.41
Non-Hindu (ref)						
Hindu	−476.66	−0.25	0.80	4.28	1.07	0.29
Backward caste (ref)						
Forward caste	1465.90	0.46	0.65	−2.81	−0.42	0.68
No education (ref)						
Primary	−508.63	−0.30	0.76	−2.54	−0.72	0.47
Middle	−970.45	−0.61	0.54	−4.53	−1.37	0.17
HS	−1192.94	−0.58	0.56	1.21	0.28	0.78
Above HS	−4935.53	−2.58	0.01	−3.55	−0.89	0.37
Kolkata (ref)						
Bhubaneswar	2885.34	1.97	0.05	20.01	6.55	0.00
Ranchi	2961.72	1.62	0.11	15.82	4.15	0.00
Intercept	62055.18	7.25	0.00	−54.53	−3.06	0.00
N	512.00			512.00		
F	10.29		0.00	10.79		0.00
R^2	0.22			0.23		

quintile group. Hindu males contribute less than non-Hindu respondents; the opposite result is obtained for female respondents. Forward caste respondents contribute less. Net contribution is relatively high in households whose main activity is casual work; it is relatively low in wage and salaried earning households.

5.6.4 Econometric Analysis

We next undertake an econometric analysis to identify determinants of (i) net contribution in absolute terms; and (ii) net contribution as percent of MPCE. An OLS model is estimated in both cases; results are reported in Table 5.11.

Results reveal that *there is no gender difference in absolute net contribution; but, in percent terms, males contribute less than female respondents*. The first OLS model (for absolute net contribution) indicates that old-old group (aged above 80 years), and those with above HS education contribute less. Lower levels of net contribution are associated with (log) MPCE. Respondents residing in Bhubaneswar also tend to contribute more. In the second model (for net contribution in percent terms), we find that net contribution is higher in larger families, and in Bhubaneswar and Ranchi.

5.7 Summing Up

In this chapter, we have tried to arrive at an estimate of the contribution of the aged. Such contribution may be divided into the following components: (i) earnings from labour markets, and (ii) imputed value of household services provided by aged. Even with the tight conditions existing in the labour market, it is not surprising that most of the aged, particularly female respondents, do not work. Those who do work, however, earn a quite high level (–3374). The main contribution of the aged, however, lies in the household services that they provide. It can either increase labour supply of earning members—or, even enable them to enter the labour market—or enable households to save on paid labour. Although women contribute more than male respondents in terms of number of days contributed to household services, their gross contribution (sum of market and non-market-based contributions) is lower than that of males. It reflects the gender discrimination in the labour market, with imputed returns of female workers being lower than that of male workers. Analysis also revealed that households spend less on females. As a result, although net contribution of the aged is negative, it is lower for female respondents.

References

Aassve, Arnstein, Bruno Arpino, and Alice Goisis. 2011. Grandparenting and mothers' labour force participation: A comparative analysis using the generations and gender survey. *Demographic Research* 27: 53–84. https://doi.org/10.4054/DemRes.2012.27.3.

Alam, M., K. S. James, G. Giridhar, K. M. Sathyanarayana, S. Kumar, S. S. Raju, T. S. Syamala, L. Subaiya, and D. W. Bansod. 2011. Report on the status of elderly in select states of India, 2011. New Delhi: UNFPA.

Albuquerque, Paula, and José Passos. 2012. Grandparents and women's participation in the labor market. In *Gender, policies and population: Conference of European Association for Population Studies*. Stockholm, Sweden.

Andrews, G.R., and M.M. Hennink. 1992. The circumstances and contributions of older persons in three Asian countries: Preliminary results of a cross-national study. *Asia-Pacific Population Journal* 7: 127–146.

Barrett, Alasdair. 2013. *The economic contribution of older Londoners*. London: Greater London Authority.

Bengtson, Vern L. 2001. Beyond the nuclear family: The increasing importance of multigenerational bonds. *Journal of Marriage and Family* 63: 1–16. https://doi.org/10.1111/j.1741-3737.2001.00001.x.

References

Del Boca, Daniela, Silvia Pasqua, and Chiara Pronzato. 2005. Fertility and Employment in Italy, France, and the UK. *Labour* 19: 51–77. https://doi.org/10.1111/j.1467-9914.2005.00323.x. Wiley/Blackwell (10.1111).

Gray, Anne. 2005. The changing availability of grandparents as carers and its implications for childcare policy in the UK. *Journal of Social Policy* 34: 557–577. https://doi.org/10.1017/S0047279405009153. Cambridge University Press.

Griggs, Julia, Jo-Pei Tan, Ann Buchanan, Shalhevet Attar-Schwartz, and Eirini Flouri. 2010. 'They've always been there for me': Grandparental involvement and child well-being. *Children & Society* 24: 200–214. https://doi.org/10.1111/j.1099-0860.2009.00215.x. Wiley/Blackwell (10.1111).

Hermalin, Albert, Carol Roan, and Aurora Perez. 1998. *The emerging role of grandparents in Asia (WP 98–52)*. Michigan.

Husain, Z., and M. Dutta. 2015. Grandparental childcare and labour market participation of mothers in India. *Economic and Political Weekly* 50: 74–82.

Husain, Zakir, and Mousumi Dutta. 2013. *Women workers in Kolkata's IT sector: Satisficing work and household*, SpringerBriefs in sociology. New Delhi: Springer. https://doi.org/10.1007/978-81-322-1593-6.

Leibowitz, Arleen, Jacob Alex Klerman, and Linda J. Waite. 1992. Employment of new mothers and child care choice: Differences by children's age. *The Journal of Human Resources* 27: 112–133. https://doi.org/10.2307/145914. University of Wisconsin Press.

Posadas, Josefina, and Marian Vidal-Fernandez. 2013. Grandparents' childcare and female labor force participation. *IZA Journal of Labor Policy* 2. 14. doi:https://doi.org/10.1186/2193-9004-2-14. SpringerOpen

Randel, Judith, Tony German, Deborah Ewing, and HelpAge International. 1999. *The ageing and development report: Poverty, independence and the world's older people*. London: Earthscan Publications.

Ruiz, Sarah A., and Merril Silverstein. 2011. Relationships with grandparents and the emotional well-being of late adolescent and young adult grandchildren. *Journal of Social Issues* 63: 793–808. https://doi.org/10.1111/j.1540-4560.2007.00537.x.

Schröder-Butterfill, Elisabeth. 2003. *Pillars of the family – Support provided by the elderly in Indonesia*. Working Paper No WP303. UK.

Timonen, Virpi, Martha Doyle, and Ciara O 'Dwyer. 2009. *The role of grandparents in divorced and separated families*. Dublin.

Zamarro, Gema. 2011. *Family labor participation and child care decisions*. Working Paper WR-833. California.

Chapter 6
Daily Life of the Aged: An Analysis of Time-Use Diaries

6.1 What Are Time-Use Studies?

The study of how people spend their time constitutes a major theme of research in social science disciplines such as anthropology, sociology, economics, and social psychology (Bolger et al. 2003; Juster et al. 2003). Such studies, going back to the beginning of the twentieth century, employ a wide range of methodologies for collecting data on time use. Such methods include direct observational approaches such as the shadowing method (Quinlan 2008) to the experience sampling method that invites respondents to record their activity at random points during the day (Csikszentmihalyi and Larson 1987; Zuzanek 2013). In recent years, time-use diaries (TUD) have emerged as a reliable and accurate data collection instrument to obtain information on the activity patterns of large populations (Robinson and Godbey 1999; Michelson 2005). After the large-scale Multinational Time Budget Research Project conducted in the 1960s (Robinson et al. 1972), a considerable number of countries, from developed to developing countries, began funding national time use surveys on a regular basis.

A time-use diary (TUD) method is a sequential and comprehensive record of the activities undertaken by the persons of interest over the course of a day or days (University of Maryland 2014). It is generally recorded by the respondent herself or himself, ideally recorded during or very soon after the time of activity. The diary is divided into regular intervals, during which the participant records what activities they were undertaking for each time interval (Löfström and Palm 2008). The basic idea underlying TUDs is that our activities are not isolated; they usually follow an order and are often repetitive on a daily basis. TUDs help in identifying such routines and distinguish between activities that are in the nature of regular occurrences, or are atypical (Ainsworth 2001).

Other contextual data may also be recorded in TUDs: who is involved in the activity (with whom or for whom), the location of the activity, the duration of the activity, and whether there are also other activities being undertaken at the same time (Vrotsou et al. 2009). Sometimes, respondents may also record reasons for atypical activities (Hektner et al. 2007).

TUDs are particularly suitable for identifying recording frequently occurring activities (Fleming and Spellerberg 1999) and understanding repetitive patterns in daily lives of respondents. In comparison to direct observation, which may be obtrusive and affect the behaviour of respondents and their household members, TUDs are more likely to ensure natural behaviour and bring out true patterns and atypical behaviour (Nachmias 1981; O'Brien 2010). Studies have also reported that TUDs are less subject to social desirability and normative response errors (United Nations Economic Commission for European 2013). Another advantage of this method is that recall error is minimized. As respondents record their own activities soon after they are undertaken, there is little scope for recall bias to distort the data. It is not surprising, therefore, that methodological comparisons of TUDs and survey question estimates have indicated higher validity and reliability of TUDs (Juster et al. 2003; Kan and Pudney 2008).

However, TUDs can be a two-edged weapon. Respondents' records are accurate only if they are familiar with and understand how to record their activities (Stinson 1999). This requires that the researcher needs to explain the form completion process to each participant (Corti 1993). Among other factors that affect accuracy of the data collected are literacy and willingness to participate (Fleming and Spellerberg 1999). These are key factors in determining data accuracy, and—particularly in developing countries, where respondents are characterized by low levels of education and awareness and are often suspicious of the motives of researchers—the violation of these underlying conditions may result in a false picture of daily life being generated.

Among other factors that limit the use of TUDs are:

1. It is cumbersome to fill in, leading to poor response rates (Chatzitheochari et al. 2018).
2. Administrative costs, in the form of intensive post-fieldwork data editing and cleaning, are often high (Minnen et al. 2014).
3. In developing countries, respondents often perform multiple activities simultaneously. This has to be recorded carefully.
4. Social customs and taboos often prevented certain activities being recorded. This was particularly true for women.
5. Another difficulty in eliciting the right response from women respondents was men frequently offering to reply for women.
6. In developing countries, particularly in rural areas, respondents do not record time, except in a broad sense. It is not possible, therefore, to measure the exact time at which the different activities occur or the time taken by people in performing the activity.

6.2 Studies on Time Use

TUDs have been used in a large number of research studies, to study different patterns of activities by different groups of population. For instance, Prodromídis (2014) has used evidence from TUDs to study the allocation of time by women on paid activities, unpaid work, and non-work in Great Britain. Fisher et al. (2007) has examined historical shifts in Americans' use of time, particularly focusing on gendered change in paid and unpaid work using TUDs. Allocation of time by infants and adolescents has been studied by Harrison et al. (2014) and Hunt and McKay (2015), respectively. TUDs have also been used in studies on health; for instance, in analysing activity patterns of women with long-term pain (Liedberg et al. 2009) and schizophrenics (Bejerholm and Eklund 2008). The impact of Internet use on daily living has been studied by Anderson and Tracey (2002), while Anderson analyses use of laundry and energy demand (Anderson 2016).

The use of TUDs is less common in India. A few studies have used small-scale time-use surveys covering a small number of villages and households in the 1980s and 1990s (Hirway 1998). The prominent studies among these are:

1. Time Allocation Study in some villages of Rajasthan and West Bengal by Jain and Chand (Jain and Chand 1982),
2. Time Allocation study in Tamil Nadu by Directorate of Economics and Statistics, Tamil Nadu (1996),
3. Time Use study by NCAER in a few villages in the 1980s, and
4. A study on the Time Use of children by Ramesh Kanbargi in Karnataka (1990).

In spite of their experimentary nature and methodological defects, these studies presented important and interesting results. The study by Jain and Chand, for instance, reported interesting observations about the time use of women, showing that their participation in economic activities was higher than what is presented in the Census of Population and National Sample Survey statistics.

To take care of the limitations of the earlier time-use studies in India and to meet the new emerging data requirements, the Ministry of Statistics and Programme Implementation (MOSPI), Government of India, conducted a pilot time-use survey during the period July 1998 to June 1999. The study covered 18,620 households spread over Haryana, Madhya Pradesh, Gujarat, Orissa, Tamil Nadu, and Meghalaya.

The main objectives of this survey were as follows:

1. To develop a conceptual framework and a suitable methodology for designing and conducting time-use studies in India on a regular basis. Also, to evolve a methodology to estimate labour force/workforce in the country and to estimate the value of unpaid work in the economy in a satellite account.
2. To infer policy/programme implications from the analysis of the data on (a) distribution of paid and unpaid work among men and women in rural and urban areas, (b) nature of unpaid work of women including the drudgery of their work, and (c) sharing of household work by men and women for gender equity.

3. To analyse the time-use pattern of the individuals to understand the nature of their work so as to draw inference for employment and welfare programmes for them.
4. To analyse the data of the time-use pattern of the specific section of the population such as children and women to draw inferences for welfare policies for them.

    To collect and analyse the time-use pattern of people in the selected states in India in order to have a comprehensive information about the time spent by people on marketed and non-marketed economic activities covered under the 1993 SNA, non-marketed non-SNA activities covered under the General Production Boundary, and on personal care and related activities that cannot be delegated to others.
5. To use the data in generating more reliable estimates on workforce and national income as per 1993 SNA, and in computing the value of unpaid work through separate satellite account.

MOSPI subsequently commissioned another pilot time-use survey in a smaller scale in the states of Bihar and Gujarat in March 2013. The data underlines the fact that

> … women's participation in SNA work is not as low as is seen from the conventional surveys, though it is commonly observed that the WPR of women in the country is very low due to socio-cultural reasons. The Indian time-use data shows that the gap between the WPRs of men and women is not as big as revealed by LFS. The major difference between men and women's WPR is that women participate, on an average, for four hours a day or less, as compared to men who participate full-time, in SNA work, which is mostly because of their domestic responsibilities. They are constrained heavily by the high burden of unpaid work, which leaves them with less time and energy to participate in productive work in the labour market. This burden is also reflected in women's low capabilities and low human capital formation as well as low mobility in the labour market. (Hirway 2017)

We have not been able to locate any study of time use by the aged—neither in India nor abroad. In fact, Andreas Mergenthaler of the Federal Institute for Population Research, Wiesbaden, had commented during a session on 'Labor market activity of older adults and care givers' at the European Population Conference held in Mainz in 2016 on the complete non-availability of time-use studies of the aged throughout the world. This implies that the TUD-based study undertaken by us is unique, not only in India but also in a global context and addresses a major lacuna in the literature on gerontology.

6.3 Objective and Methodology

One of the main research questions of the study was to assess the extent to which the aged are integrated with their family and society at large. This indicates whether the Indian elderly are ageing actively, or not. Active ageing refers to not just the ability to be physically active or to participate in the labour force, but to continuing

6.3 Objective and Methodology

participation in social, economic, cultural, spiritual, and civic affairs (World Health Organization 2002). The premise underlying the concept of active ageing is that older people who retire from work and those who are ill or live with disabilities can continue to remain active contributors to their families, peers, communities, and nations.

Although Labour Force Surveys (LFSs) provide information on the extent to which respondents are engaged in economic activities, they suffer from certain limitations:

> These limitations have emerged from several relatively recent developments, namely: (a) changes in the production boundary of the UN System of National Accounts (SNA) that have made it difficult to capture workforce statistics adequately, (b) changes in the labour market structures and in the characteristics of workforce and labour force over the years that call for some innovative approaches to estimate and understand the flexibilities of work, and (c) the new classification of SNA work under the resolution (unpaid trainee work and voluntary work of certain kinds) that calls for modification in the workforce/labour force classifications. In addition, there is also an acceptance that the causes of the inferior status of women, as compared to men, in the labour market lie in the unequal sharing of unpaid work by men and women (Hirway 2017)

Time-use surveys (TUS), on the other hand, have the potential to address the limitations of the LFS, as they provide comprehensive information on how individuals spend their time on SNA activities, non-SNA activities that fall within the general production boundary, and personal activities.

Accordingly, the last section of the questionnaire used in our study contained a section on time-use patterns of the respondents. The activities undertaken in the last two *normal working* days were recorded from the time the respondents woke to the time she/he went to bed at night. The starting and ending time was used to estimate sleeping hours of the respondent.

In the editing stage, the activities were grouped into eight secondary activities that were then classified under four broad heads: direct economic activities (SNA activities), indirect economic activities (or extended SNA activities), leisure or recreational time spent with other people (including family members, friends, relatives, etc.), and personal time (spent alone). This classification, details of which are given in Table 6.1, enables us to assess the extent to which active ageing is taking place in India. The first three primary heads together comprise active ageing.

Activities were recorded over each hour. In case of multiple activities, the two main activities were recorded, giving each of them equal weightage (30 min, each). This reduced the tedium of minutely recording activities, by omitting activities being undertaken for less than 15 min, and ensured a good response rate. Since quite a few respondents had no or low levels of respondents, the investigators explained the concept of TUDs to respondents and filled up the schedule with their inputs.

Investigators were instructed that if any of the preceding last 2 days were atypical, in the sense that there were deaths or any festivity or any special ceremony, the day(s) before should be taken. For instance, if the interview was undertaken on Day 0, the activities should normally be recorded for Days (-1) and (-2); however, if there was a death in (say) Day(-1), then the activities should be recorded for

Table 6.1 Activities recorded and their classification

Major	Secondary	Primary
Working	Working	Economic activities
Travelling to work	Working	
Bank	Outside household chores	Extended SNA activities
Daily marketing	Outside household chores	
Marketing for groceries	Outside household chores	
Pay bills	Outside household chores	
Ration shop	Outside household chores	
Cooking	Household chores	
Dusting furniture	Household chores	
Sweeping house	Household chores	
Washing clothes	Household chores	
Playing with children	Child-related activities	
Supervise children's homework	Child-related activities	
Travelling to drop/pick up child to/from school, etc.	Child-related activities	
Morning/evening walk	Social life	Socializing
Social life	Social life	
Having a meal	Personal activities	Personal
Newspaper, TV, radio	Personal activities	
Personal care (washing, dressing, and toilet)	Personal activities	
Resting/relaxing	Personal activities	
Medical	Miscellaneous	
Other activity	Miscellaneous	
Other travel	Miscellaneous	
Sleeping	Sleep	

Day(−2) and Day(−3). The rationality of this instruction was to avoid atypical activities.

Further, the preceding 2 days should be working days, and not holidays. The reason was that a working aged person would not report his association with economic activities. Nor would activities related to dropping a child at school, or picking her/him from school, be reported. Activities like paying utility bills would also not be reported. Overall, a bias towards reporting leisure-based activities may occur if non-working days are included in the study. Again, if any of the 2 days preceding the day of survey was a working day, Day(−3) or even Day(−4) would be taken.

6.4 Main Findings

6.4.1 Mean Time Spent on Activities

The mean time spent on different activities is given in Table 6.2. While personal activities, of which sleeping is an important component, comprises the dominant activity for the aged, they are also engaged in SNA and extended—SNA activities. A gendered division may be noted here—with *men engaged in economic activities*

Table 6.2 Mean time spent on each activity (minutes)

Activities	Gender	Mean	Std. Error	95% Confidence Interval	
Sleeping (31.0***)					
	Female	567	7	552	581
	Male	536	7	522	549
Miscellaneous (−21.4**)					
	Female	233	7	218	247
	Male	254	9	237	271
Personal (−27.5**)					
	Female	346	8	330	362
	Male	319	9	302	336
Aggregate of personal activities (37.1**)					
	Female	1146	10	1126	1165
	Male	1108	12	1085	1132
Work (−74.6***)					
	Female	42	7	28	56
	Male	117	12	92	141
Childcare services (−6.4)					
	Female	30	4	22	37
	Male	36	4	28	45
Household chores (95.9***)					
	Female	110	7	96	123
	Male	14	2	9	18
Outside home chores (−31.9***)					
	Female	11	2	8	14
	Male	43	3	37	49
Extended SNA activities (57.7***)					
	Female	150	8	135	165
	Male	93	6	81	104
Social activities (−20.2**)					
	Female	102	6	90	114
	Male	122	7	109	135
Active time (−37.1**)					
	Female	294	10	275	314
	Male	332	12	308	355

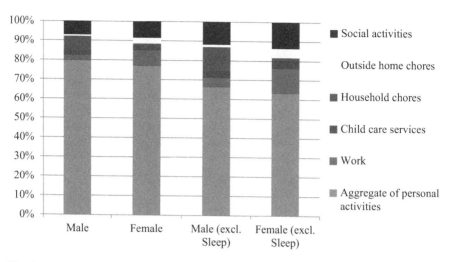

Fig. 6.1 Mean time spent on different activities as share of total day

and women participating in extended SNA activities (even here, male respondents spend more time than women on outside home extended-SNA activities). Similarly, women respondents are engaged more in personal activities, while men tend to socialize more significantly. In all, *men tend to age more actively than women.*

This is also confirmed if we examine Fig. 6.1 where the allocation of time across different activities is given. Since sleep is an integral activity, we have constructed two bars for each gender. The first one includes sleeping time (i.e. shows the allocation for the entire day), while the other excludes sleeping time (i.e. shows the allocation of time across 24 h less sleeping time).

6.4.2 Percentage Time Allocated on Broad Activity Groups

The activities studied in the earlier section are too broad for meaningful analysis. We have, therefore, clubbed them into broad activity groups as shown in Table 6.1. In Fig. 6.2, we present estimates of time allocated (as percentage of waking time) between the broad activity groups comprising active ageing: work, extended SNA activity, and socializing. We also present the aggregate of time allocated on all these activities. This comprises active ageing.

Predictably, the aged male respondents spend a greater time in economic activities, compared to female respondents. In case of extended SNA activities, the opposite is observed.[1] What is interesting is that *if we compare the aggregate time allocated on economic and extended-SNA activities together, the total comes to 22.7*

---

[1] Both differences are statistically significant: $t = 5.2672$ and $= -4.4728$, respectively, for economic activities and extended SNA activities, with $p = 0.0$ in both cases.

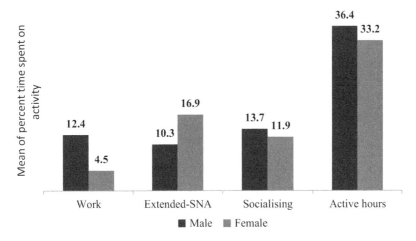

Fig. 6.2 Percentage of time spent on broad activity group

and 21.3 percent for male and female respondents, respectively. The marginal difference[2] indicates that women contribute an equal number of hours to working; however, the fact that these hours are spent on household activities implies that females remain under-valued and under-recognized contributors to family welfare (Randel et al. 1999; WRVS 2011). Results also reveal that elderly men tend to socialize significantly more than their female respondents.[3] In all, respondents spend about one-third of their waking time actively, with male respondents being significantly more active.[4] In the absence of alternative estimates, it is not possible to conclude whether India is ageing actively; however, one-third appears somewhat on the lower side.

6.4.3 Variations in Percentage Time Allocated on Broad Activity Groups Across Correlates

We next examine variations in the percentage time allocated to different broad activity groups across socio-economic correlates. This is given in Table 6.3.

It can be seen that among male respondents, those from self-employed households spend about a third of their waking time in working and almost half their waking time actively engaged. Among female respondents, those from regular wage and salaried households spend about a fourth of their waking time in extended SNA

---

[2] The difference is also statistically insignificant, as $t = 0.7939$ ($p = 0.4276$).

[3] The difference is statistically significant, with $t = 1.7138$ $9p = 0.0436$).

[4] Given $t = 1.9538$($p = 0.0256$) the difference is again statistically significant.

Table 6.3 Variations in proportion of time allocated to broad activity group

Correlates	Working		Extended SNA		Socializing		Active ageing	
	Male	Female	Male	Female	Male	Female	Male	Female
Household type								
Self-employed	34.0	9.4	7.7	19.2	8.2	8.2	49.9	36.9
Regular wage earner	9.1	2.6	9.4	23.0	15.8	5.9	34.3	31.5
Casual labour	22.5	19.8	7.8	15.1	6.9	8.4	37.2	43.4
Others	4.0	2.7	12.1	12.9	16.2	16.7	32.2	32.2
Expenditure quintile								
Poorest	9.5	1.8	8.9	15.0	17.3	19.8	35.7	36.6
Low middle	15.6	11.6	10.5	14.3	13.1	11.8	39.3	37.6
Middle	7.4	4.9	13.0	13.5	15.1	10.1	35.4	28.5
High middle	13.4	3.3	11.0	22.5	11.8	7.6	36.3	33.4
High	17.8	2.2	6.8	18.2	11.0	8.0	35.7	28.4
Education								
No education	14.8	5.8	15.2	11.7	11.4	16.9	41.3	34.5
Primary or below	19.1	8.6	6.9	14.4	9.4	11.8	35.4	34.8
Middle and secondary	13.2	3.4	10.8	19.3	13.3	10.8	37.3	33.5
Higher secondary	13.9	4.3	9.7	21.7	15.7	8.1	39.3	34.2
Above HS	6.3	0.5	9.7	23.3	16.2	5.5	32.2	29.3
Age groups								
60–65 years	18.2	6.2	10.1	20.3	12.6	11.8	40.9	38.3
66–70 years	7.5	3.2	11.7	17.1	12.5	12.9	31.7	33.2
71–80 years	7.6	3.4	9.7	14.6	15.0	11.3	32.3	29.3
81 years and above	1.1	1.2	10.2	5.4	19.6	11.5	30.8	18.0
Religion								
Non-Hindu	11.9	7.5	11.7	14.8	13.3	14.2	37.0	36.5
Hindu	12.4	3.8	10.2	17.3	13.7	11.4	36.3	32.5
Caste								
Backward castes	12.3	4.6	10.3	17.1	13.7	11.6	36.2	33.3
Forward castes	17.6	1.1	11.3	11.4	15.5	19.4	44.3	31.9
City								
Kolkata	8.2	1.5	9.8	22.0	17.6	7.6	35.6	31.0
Bhubaneswar	14.3	5.2	10.4	10.0	10.7	19.0	35.5	34.2
Ranchi	18.0	9.6	11.2	19.1	11.0	7.4	40.2	36.2

activities, while more than two-fifth of the waking time of females from casual labourer households is spent actively.

The wealth-time use gradient is not always regular. It displays an inverse U-shape relationship for extended SNA activities among male respondents and economic activities among female respondents; the gradient is negative for socializing activities among both male and female respondents. A negative trend is also observed among female respondents for active ageing. In other cases, no clear trend is apparent.

In case of education, again, there is absence of clear relationship. Among male respondents, the education–time-use gradient is inverse U shape for economic activities but U-shaped for socializing. Among female respondents, the education–time use is positive for extended SNA activities but negative for socializing. Among female respondents, the highly educated segment (with above HS education) is least active.

As respondents age, they devote less time to economic activities, extended activities, and in active ageing, in general. This is observed for both male and female respondents.

6.5 Econometric Analysis

6.5.1 Active Ageing as Dependent Variable: Ordinary Least Squares Model

The dependent variable of the econometric analysis is percentage hours spent actively. This is a continuous variable, and we can use an OLS to estimate this model. Results of the OLS model are given in Table 6.4. Tests for multi-collinearity and heteroskedasticity are also reported. While there is no multi-collinearity (Variance Inflation Factor < 10, the critical value indicating multicollinearity), tests for heteroskedasticity indicate that the assumption of constant variance does not hold. So, we estimated robust standard errors.

It may be seen that women respondents spend less time actively, compared to their male counterparts. Similarly, as a person ages, she/he becomes less active. Neither religion nor caste appears to have any effect. Interestingly, it is the affluent respondents who are ageing actively. This would imply that it is economic pressures that force the less affluent aged respondents to work hard. However, Table 6.3 shows that this is not necessarily true. While the low- to middle-income group does work more, so does the high- to middle-income group. Moreover, the poorest group spends almost one-fifth of their waking times socializing. So, it is not economic pressures, but possibly the fact that poorer households live in smaller accommodation and are more closely knit than richer families.

Active ageing does not vary across education levels; we find that only those with graduate, postgraduate, professional, or higher levels of education age less actively than less educated respondents.

Finally, residents of Ranchi age more actively than residents of Kolkata and Bhubaneswar.

Table 6.4 Results of OLS model

Variables	Coefficient	t	P > t
Male (ref)			
Female	−4.22	−2.63	0.01
Non-Hindu (ref)			
Hindu	0.41	0.14	0.89
Backward castes (ref)			
Forward caste	−0.34	−0.08	0.93
Age	−0.66	−4.25	0.00
Log(MPCE)	−2.52	−2.72	0.01
No education (ref)			
Primary	−0.98	−0.34	0.73
Middle	−1.55	−0.59	0.56
Higher secondary	−0.12	−0.04	0.97
Above HS	−6.59	−2.28	0.02
Kolkata (ref)			
Bhubaneswar	−1.30	−0.59	0.56
Ranchi	5.06	2.09	0.04
Intercept	118.34	7.11	0.00
N	513		
R^2	0.12		
F	4.53		0.00
VIF	1.70		
Breusch-Pagan statistic	12.72		0.00

6.5.2 An Alternative Econometric Model

Now, there is one possible methodological issue with the choice of OLS model. The values of the dependent variable are not free to vary and assume any value. They will range from 0% to 100% as a person cannot spend less than 0 h actively, not spend more than 24 h actively. The actual distribution of the variable 'PACTIVE' is from 0 to 87.5, with 25 observations taking the minimum value of 0 (about 4.8%). The kernel density is given in Fig. 6.3.

The distribution is roughly normal, with a probability spike at zero (Fig. 6.4). Tests for normal distribution indicate that the distribution may be considered to be normal: the Doornik–Hansen χ^2 multivariate normality statistic is 3.79 ($p = 0.15$); the Shapiro–Wilks test also indicates normality ($W = 0.99$, $p = 0.00$). In such cases, the implication of zeroes has to be examined carefully. If, for instance, negative values of PACTIVE are reported as zero, then we have a censored distribution, so that a Tobit model is appropriate. However, in our case, given that the probability spike is not very large (less than 5%), we would not expect the two models to differ substantially.

6.5 Econometric Analysis

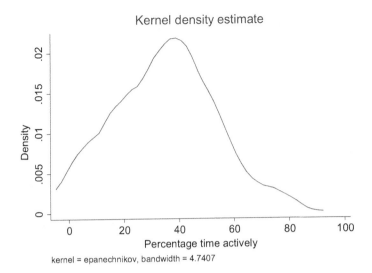

Fig. 6.3 Kernel density of percentage hours spent actively

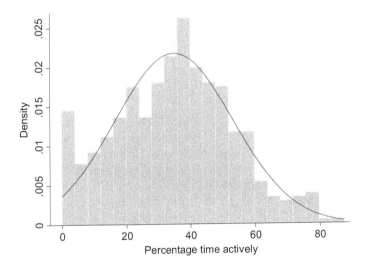

Fig. 6.4 Histogram of percentage hours spent actively

A Tobit model, therefore, is also estimated and reported (Table 6.5). As expected, the results do not vary substantially. Variables that are (in)significant in the OLS model remain so in the Tobit model; there is only a marginal difference between the estimated coefficients.

Table 6.5 Results of Tobit model

Variables	Coefficient	t	$P > t$
Male (ref)			
Female	−4.25	−2.47	0.01
Non-Hindu (ref)			
Hindu	0.18	0.06	0.95
Backward castes (ref)			
Forward caste	−0.67	−0.13	0.90
Age	−0.70	−6.99	0.00
Log(MPCE)	−2.63	−2.50	0.01
No education (ref)			
Primary	−1.09	−0.39	0.70
Middle	−1.48	−0.56	0.57
Higher secondary	0.23	0.07	0.95
Above higher secondary	−6.65	−2.11	0.04
Kolkata (ref)			
Bhubaneswar	−1.22	−0.52	0.60
Ranchi	5.48	1.93	0.05
Intercept	122.67	7.68	0.00
N	513		
Censored observations	25		
Pseudo-R^2	0.02		
χ^2	68.62		0.00

6.6 Conclusion

To sum up, analysis of time-use data reveals that respondents spend about one-third of their waking time actively. This seems on the lower side. Gender differences are noted—both in active ageing and in the allocation of waking time. Men spend more time on socializing, on economic activities, and on extended SNA activities performed outside the home. On the other hand, women tend to spend more time on household chores like cooking, cleaning the house, and on personal activities. This indicates that the norms of a highly gendered patriarchal society that limit women to performing certain activities within the physical domain of the house continue to operate on the aged. Women remain confined within the house, interacting with family members, undertaking household chores, and on personal activities. Such activities are little valued socially, so that the contribution of women to the household (in the form of household chores) are not recognized by their family members, or even themselves. This is a concerning finding and requires the attention of policymakers.

References

Ainsworth, Susan. 2001. Discourse analysis as social construction : Towards greater integration of approaches and methods. In *Second international critical management studies conference*, 26. Julio 11–13. Manchester School of Management. UMIST: UK.

Anderson, Ben. 2016. Laundry, energy and time: Insights from 20 years of time-use diary data in the United Kingdom. *Energy Research & Social Science* 22: 125–136. https://doi.org/10.1016/J.ERSS.2016.09.004. Elsevier.

Anderson, Ben, and Karina Tracey. 2002. Digital living: The impact (or otherwise) of the internet on everyday British life. In *The internet in everyday life*, 139–163. Oxford: Blackwell Publishers Ltd. https://doi.org/10.1002/9780470774298.ch4.

Bejerholm, Ulrika, and Mona Eklund. 2008. Time use and occupational performance among persons with schizophrenia. *Occupational Therapy in Mental Health* 20: 27–47. https://doi.org/10.1300/J004v20n01_02. Taylor & Francis Group.

Bolger, Niall, Angelina Davis, and Eshkol Rafaeli. 2003. Diary methods: Capturing life as it is lived. *Annual Review of Psychology* 54: 579–616. https://doi.org/10.1146/annurev.psych.54.101601.145030. Annual Reviews 4139 El Camino Way, P.O. Box 10139, Palo Alto, CA 94303–0139, USA.

Chatzitheochari, Stella, Kimberly Fisher, Emily Gilbert, Lisa Calderwood, Tom Huskinson, Andrew Cleary, and Jonathan Gershuny. 2018. Using new technologies for time diary data collection: Instrument design and data quality findings from a mixed-mode pilot survey. *Social Indicators Research* 137: 379–390. https://doi.org/10.1007/s11205-017-1569-5. Springer Netherlands.

Corti, Louise. 1993. Using diaries in social research. *Social Research Update*.

Csikszentmihalyi, M., and R. Larson. 1987. Validity and reliability of the experience-sampling method. *The Journal of Nervous and Mental Disease* 175: 526–536.

Fisher, Kimberly, Muriel Egerton, Jonathan I. Gershuny, and John P. Robinson. 2007. Gender convergence in the Americal Heritage Time Use Study (AHTUS). *Social Indicators Research* 82: 1–33. https://doi.org/10.1007/s11205-006-9017-y. NIH Public Access.

Fleming, Robin, and A. Spellerberg. 1999. *Using time use data. A history of time use surveys and uses of time use data*. Wellington: Statistics New Zealand.

Harrison, Linda J., Sheena Elwick, Claire D. Vallotton, and Gregor Kappler. 2014. Spending time with others: A time-use diary for infant-toddler child care. In *Lived spaces of infant-toddler education and care*, 59–74. Dordrecht: Springer. https://doi.org/10.1007/978-94-017-8838-0_5.

Hektner, Joel M., Jennifer A. Schmidt, and Csikszentmihalyi Mihaly. 2007. *Experience sampling method: Measuring the quality of everyday life*. Thousand Oaks: Sage.

Hirway, Indira. 1998. Time use studies: Conceptual and methodological issues with reference to the Indian time use survey.

———. 2017. *India's time use data show women spend 69.03 hours on total work, men spend 62.71 hours*. Blog posted on 17 April. Accessed on 6 May 2018 from https://bit.ly/2ICeRyQ.

Hunt, Eithne, and Elizabeth Anne McKay. 2015. What can be learned from adolescent time diary research. *Journal of Adolescent Health* 56: 259–266. https://doi.org/10.1016/j.jadohealth.2014.11.007.

Jain, Devaki, and Malini Chand. 1982. Report on a time allocation study – Its methodological implications. In *Technical seminar on 'Women's Work and Employment'*. New Delhi: Institute of Social Studies Trust.

Juster, F. Thomas, Hiromi Ono, and Frank P. Stafford. 2003. An assessment of alternative measures of time use. *Sociological Methodology* 33. 19–54. https://doi.org/10.1111/j.0081-1750.2003.t01-1-00126.x. Wiley/Blackwell (10.1111)

Kan, Man Yee, and Stephen Pudney. 2008. Measurement error in stylized and diary data on time use. *Sociological Methodology* 38: 101–132. https://doi.org/10.1111/j.1467-9531.2008.00197.x. Wiley/Blackwell (10.1111).

Liedberg, Gunilla M., Malin E. Hesselstrand, and Chris M. Henriksson. 2009. Time use and activity patterns in women with long-term pain. *Scandinavian Journal of Occupational Therapy* 11: 26–35. https://doi.org/10.1080/11038120410019081.

Löfström, Erica, and Jenny Palm. 2008. Visualising household energy use in the interest of developing sustainable energy systems. *Housing Studies* 23: 935–940. https://doi.org/10.1080/02673030802425602. Taylor & Francis Group.

Michelson, William. 2005. Time use: Expanding the explanatory power of the social sciences. *Contemporary Sociology: A Journal of Reviews* 36: 284–285.

Minnen, Joeri, Ignace Glorieux, Theun Van Tienoven, S. Daniels, D. Weenas, and J. Deyaert. 2014. MOTUS: Modular online time-use survey-translating an existing method in the 21st century. *Electronic International Journal of Time Use Research* 11: 73–93.

Nachmias, David. 1981. *Research methods in the social sciences*. New York: St. Martin's Press.

O'Brien, Michelle. 2010. *Unobtrusive research methods – An interpretative essay*.

Prodromídis, Pródromos-Ioánnis K. 2014. Approaching the female labor supply from the unpaid work and non-work functions: Time-use diary evidence from Britain, 1998–9. *International Journal of Manpower* 35: 643–670. https://doi.org/10.1108/IJM-05-2014-0121. Emerald Group Publishing Limited.

Quinlan, Elizabeth. 2008. Conspicuous invisibility. *Qualitative Inquiry* 14: 1480–1499. https://doi.org/10.1177/1077800408318318. SAGE Publications Sage CA: Los Angeles, CA.

Randel, Judith, Tony German, Deborah Ewing, and HelpAge International. 1999. *The ageing and development report: Poverty, independence and the world's older people*. London: Earthscan Publications.

Robinson, John P., and Geoffrey Godbey. 1999. *Time for life: The surprising ways Americans use their time*. Vol. 104. University Park: Pennsylvania State University Press. https://doi.org/10.1086/210016.

Robinson, John P., Philip E. Converse, and Alexander Szalai. 1972. *The use of time: Daily activities of urban and suburban populations in twelve countries*. The Hague: Mouton.

Stinson, Linda L. 1999. Measuring how people spend their time: A time-use survey design. *Monthly Labor Review* 122: 12. https://doi.org/10.1086/250095.

United Nations Economic Commission for European. 2013. *Harmonised European time use surveys*.

University of Maryland. 2014. Time use laboratory. Measuring time use.

Vrotsou, Katerina, Kajsa Ellegård, Matthew Cooper, Katerina Vrotsou, Kajsa Ellegård, and Matthew Cooper. 2009. Exploring time diaries using semi-automated activity pattern extraction. *electronic International. Journal of Time Use Research* 6: 1–25. Research Institute on Professions (Forschungsinstitut Freie Berufe (FFB)) and The International Association for Time Use Research (IATUR).

World Health Organization. 2002. *Active ageing: A policy framework*. Geneva: WHO.

WRVS. 2011. *Valuing the socio-economic contribution of older people in the UK*. Cardiff: WRVS.

Zuzanek, Jiri. 2013. Does being well-off make us happier? Problems of measurement. *Journal of Happiness Studies* 14: 795–815. https://doi.org/10.1007/s10902-012-9356-0.

Chapter 7
Conclusion

7.1 Returning to Research Questions

The primary research question of our study is whether the elderly are an asset to their family or a burden? To facilitate analysis, the question was broken up into the following sub-questions:

1. What is the economic contribution of the elderly?
2. What is the non-market contribution of the elderly?
3. What is the gross economic contribution of the elderly?
4. What is the net economic contribution of the elderly?
5. What is the time use pattern of aged? How much of it is spent on contributing to the family?
6. Does the aged in India age actively?

We briefly sum up the light thrown by our research on these questions.

7.2 Main Findings

7.2.1 Economic Contribution of the Elderly

Information on earnings of workers is obtained from the current weekly status data of National Sample Survey Office (NSSO). Using this data it is possible to calculate daily earnings of workers. Daily earnings are multiplied by 30 to arrive at monthly earnings. However, NSSO does not provide the earnings of informal sector workers either as own account workers or as workers in household enterprises. Not considering earnings of elderly in informal sector underestimates the true contribution of them, as the informal sector is the largest employment providing sector in India. So, in this chapter, the monthly earnings of informal sector workers including self-employed person are estimated.

The monthly earnings of the informal sector workers in the household are obtained by considering households with workers in the informal sector and subtracting earnings of regular wage and salary earners and casual workers from the total monthly earnings of each household. The logic is that households in the informal sector are unlikely to have substantial savings—so the 'extra' consumption must have been financed by the earning from the informal sector.

To obtain the earnings data of the elderly people in the informal sector, we assume that both elderly and non-elderly workers secure work for the same number of days. Dividing the informal sector earnings with the total workers in the informal sector, we arrive at average monthly earnings of informal sector workers in the households and then this is multiplied by a number of elderly workers in each household in the informal sector to get informal sector earnings of aged workers.

The aggregate monthly earnings of all aged workers (including earnings of informal sector workers) in each household is used to estimate the gross financial contribution of the aged. This is defined as follows:

$$\text{Gross financial contribution of aged} = 100 \times \frac{(\text{Total monthly earnings of all aged workers})}{\text{Total monthly household expenditure}}$$

Using the above formula, we estimate mean gross contribution. The above formula gives some idea of the financial contribution of the elderly to the household. Our analysis revealed that gross financial contribution was about the same (19%) in both the rounds. Substantial rural-urban differences were observed. Gross financial contribution of aged is higher in rural areas (21 and 24% in the two rounds), as compared to that in urban areas (14 and 15%). A positive gross contribution is also observed in most of the Indian states; t-tests indicate that such contributions differ significantly from zero in most of the states in both urban and rural states.

However, a more appropriate indicator to capture the actual contribution of elderly to the household should be one that deducts, from gross contribution, the amount of money income spent for their (elderly) own expenses. So we have focussed on the net contribution of elderly, given as follows:

Net financial contribution of aged =
$$100 \times \frac{(\text{Total monthly earnings of all aged workers} - \text{Total monthly expenditure on aged})}{\text{Total monthly household expenditure}}$$

Net contribution helps us to analyse whether the elderly are contributing more to the family than their share of consumption expenditure.

Monthly per capita expenditure (MPCE) of the household is considered as a proxy of monthly per capita income of the household. We have also assumed that both elderly and non-elderly have same per capita monthly expenditure. The pattern of expenditure of the elderly differs from other age groups, for instance, health

7.2 Main Findings

expenditure may be high for the elderly. But per capita expenditure may not differ significantly as other components like food, transport, and recreation may be lower for the aged. Multiplying monthly per capita expenditure with the total household members, we arrive at a total monthly expenditure of the household.

Our analysis reveals that in about 32% of rural households, the aged make a positive contribution; in urban areas, in about one out of every five households, the aged make a positive contribution. The proportion is about the same in both rounds. In rural and urban India, in both the rounds, the poorest elderly are found to be contributing more towards their family expenditure than aged from more affluent households. The relatively high contribution of the lowest expenditure group of elderly implies that economic pressure and need for survival are major driving forces behind the participation in the work and to contribute a larger portion of their income to household expenditure. Lower contribution of the higher expenditure group may be because of the better-off economic condition of the younger household members or because the aged have saved during their working period.

Econometric analysis to identify determinants of net financial contribution in rural areas reveals that household expenditure level does not have any significant relation with a net contribution of rural elderly in both the rounds. Predictably, rural elderly workers contribute more in larger families and in families with a smaller number of non-aged working members in both the rounds. Higher state-level unemployment rates are negatively associated with lower net financial contribution of the elderly in both the rounds. In rural India, per capita net state domestic product does not show any significant relationship with the net contribution of elderly in the 55th round of NSS. However, in the 68th round, we have observed a negative relationship between the log of net state domestic product and net contribution of elderly. Significant variations in the magnitude of net financial contribution are observed across socio-religious groups in the 55th and the 68th round of NSS in rural India. In the 55th round, we have found that Hindu Scheduled Tribes households are more dependent on the aged compared to Hindu Upper Caste families. In the 68th round, not only Hindu Scheduled Tribes households but also Muslim, Hindu Scheduled Caste, and Hindu Other Backward Caste households are more dependent on aged than the Hindu Upper Caste families.

In urban India, in both the rounds, household monthly per capita expenditure does not show any significant relation with a net contribution of elderly. In urban India, elderly are contributing more in larger families and in families with a smaller number of non-aged working members in both the rounds of NSS. Per capita net state domestic product shows significant relation only in the 68th round. Higher the per capita net state domestic product, lower the contribution of the elderly has been observed in urban India in the 68th round. In the 68th round, growth may have to make the household prosperous which reduces the pressure on the aged to participate in work. The coefficient of state-level unemployment is significant and negative in both the rounds. Among different socio-religious groups, Muslim and Hindu Other Backward Caste elderly are contributing significantly more than the Hindu Upper Caste elderly in 1999–2000 and 2011–2012. Hindu Scheduled Tribes elderly only in the 68th round are contributing significantly higher than the Hindu Upper Caste elderly. However, the contribution of Hindu Scheduled Caste elderly is not significantly different from the Hindu Upper Caste in 2011–2012.

In the next stage of our analysis we assessed the contribution of the aged in reducing incidence and intensity of poverty. The head count ratio (HCR) and Foster Greer Thorbecke (FGT) Index are used in our analysis. In the first step we estimated HCR and FGT using Planning Commission poverty lines for both rounds and for both rural and urban areas. In the second step we deducted earnings of the aged from total household consumption and divided the resultant figure by number of non-aged household members. It gives us the monthly per capita expenditure in the absence of aged members. Both HCR and FGT were recalculated using the revised monthly per capita expenditure estimates.

It was found that both HCR and FGT increased substantially if we dropped aged members. In rural areas, HCR increased by 8.5 and 8.9% (55th and 68th rounds, respectively); in urban areas, the increase was 5.0 and 6.2% (55th and 68th rounds, respectively). The increase in intensity of poverty was even higher, about 80% in rural areas and about 90% in urban areas. While regional variation was observed, in general, the intensity and incidence of poverty as found to have increased in both rounds and in both rural and urban areas in the majority of states.

7.2.2 Economic Contribution of the Elderly: A Regional Profile

Analysis of the data from the primary survey revealed that 17.9% of the respondents were working, while 5.4% were not working but seeking work (unemployed). Workforce participation was higher among males (26.5%), relative to that among females (9.7%). Most of the working respondents are manual workers, engaged in retail trade or in the residual category 'Others'. The average monthly earnings from work is Rs. 3374; it is Rs. 4182 for male workers, while for female workers, the average is much lower (Rs. 1429). Most respondents, particularly females, do not retain this earning but contribute it to the female kitty.

The monthly earnings comprise gross financial contribution of the elderly to their families. If we consider all families, the average gross financial contribution is 4.8%; it is 5.7% for male respondents and 4.4% for female respondents. If we consider only families where the aged members are working, gross financial contribution rises marginally to 6.5% and 5.1% for male and female respondents, respectively. Elderly persons contribute more in low-income households, households engaged in either informal sector or casual work, and in Ranchi. Analysis also reveals that respondents without education or respondents with HS or higher levels of education, non-Hindus, those belonging to SCs/STs, and those in the 60–65 or 71–80 year age groups contribute more than other categories.

7.2 Main Findings

We have used a regression model to identify determinants of earnings. Given that there is non-random truncation of the sample, with a group not working and so not earning either, use of a Heckman model is recommended. Household size is used as the instrument. Results show that *male respondents have a higher probability of working; however, there is no gender gap in earnings.* Age does not have an effect on either probability of working (except that those aged 71–80 years are less likely to work); nor does it affect the level of earnings. Coefficients of education, religion, caste, and household expenditure level are found to be insignificant even at 10% level. Respondents from Ranchi are found to have a higher probability of working and also earning more than respondents from other cities.

7.2.3 Indirect Contribution of the Elderly: A Regional Profile

Although aged respondents are found to supplement household income by working beyond their age of superannuation, this is not their main contribution. Elderly people are often engaged in extended SNA (System of National Accounts) activities, performing household chores. This has two effects—first, it frees other members (particularly females) who can enter the labour market and supplement household income; second, it increases household disposable income by saving the expenditure on paid workers, who would otherwise have to be hired to undertake the household chores.

Analysis of the mean number of days that respondents perform household chores reveals that *women contribute more than male respondents in terms of number of days*. It may be attributed to the *gendered division of household chores*: males generally undertake financial and activities outside the physical space of the household, while females perform chores within the living space of the family. Further, activities like cooking, washing clothes, and cleaning the house are daily activities; in comparison, male activities like marketing, paying utility bills, performing financial services, and queuing in ration shops are performed at intervals. As a result, female respondents work for a larger number of days than male respondents.

Analysis of data also reveals that the aged serve food to their children, take them to the doctor or nurse them (when sick), play with them, and watch TV along with them. There is a gendered division of labour with respect to medical services, with grandfathers taking the child to the doctor, but grandmothers nursing them. A large proportion of grandmothers also serve food to the children. Both grandparents, however, play with the children, or accompany them in watching TV. Interestingly, the proportion of aged taking children to schools, or bringing them back, is quite low—only about one out of every five respondents report undertaking such chores. Nor do grandparents take the children out for extra-curricular activities or supervise their homework.

While studies have reported such indirect economic contributions of the aged to their families, we have not been able to locate any study that has tried to estimate the economic value of such contributions. This is a major lacuna of studies on gerontology, given that such indirect economic contributions can form a major share of total contribution of aged to their families. In this study, we have attempted to estimate the indirect economic contribution of the aged in money terms. This is not easy as there is no market for unpaid family labour, so that we can use wages or earnings to value such labour. The method employed for estimating the economic value of unpaid family labour is described below.

Our starting proposition is that the economic benefit of the aged performing a chore differs between a low-income and high-income household. In case of low-income households, when an aged member performs a household chore, this releases the female member who can seek paid work. The supplemental income is the value of non-economic contribution of the aged. On the other hand, for middle- and high-income households, the help of aged family members will allow the family to reduce dependence on paid labour. The savings from this comprises the indirect economic contribution of the aged. Therefore, the value imputed to indirect economic contribution of the aged is the potential earnings of female member (in case of low-income households), or the avoided expenditure on paid providers of household services. The daily value of this imputed value is multiplied by x (when $x = 30$/number of days aged provided the service). While taking the imputed value the following points were taken into account:

(a) *Type of service provided by aged*: For instance, if the respondent paid utility bills, then the income of a driver was used for the imputation. Similarly, if the respondent, cooked, the payment to a maid servant was used.
(b) *Economic status of household*: Payment to hired help will depend upon economic status of households. So we used different rates for the same service for middle- and high-income households.
(c) *City in which respondent is located*: Costs are higher in Kolkata. We used average costs in Kolkata for Kolkata respondents, and half of this for respondents in other cities.

The mean imputed value of household services provided by the aged is Rs. 2037. It is interesting to note that this value is higher for male respondents (Rs. 2116, against Rs. 1919 for females). The reason lies in the gender discrimination present in the labour market; the payment for labour services provided by female workers is generally lower than that of male workers, so savings from household chores performed by female respondents is lower than that of male respondents.

7.2.4 Gross Economic Contribution of the Elderly

We have estimated two models to identify determinants of gross financial contribution. The first model is estimated using the Instrument variable model as gross contribution and monthly household expenditure may have a two-way causality. Again,

7.2 Main Findings

household size is used as an instrument. The Wu-Hausman test supports our hypothesis of endogeneity; the F-test indicates that the instrument used (household size) is relevant—strongly correlated with monthly household expenditure. In addition, we also used a Tobit model, with lower limit of zero, to identify determinants of gross contribution. Both models are qualitatively similar so that results are robust. Results indicate that monthly household expenditure is positively related with gross contribution; the coefficient is significant at 1% level. *Male respondents make a higher economic contribution to their families.* Compared to respondents with no education or those with primary education, respondents with middle level of education or with above school education contribute less to households. Aged from Hindu households contribute less than their non-Hindu counterparts; caste, however, does not have any impact.

7.2.5 Net Economic Contribution of the Elderly

Based on the imputation we obtained a money value for the unpaid family labour provided by the aged. This is used to estimate net contribution of aged as follows:

$$\frac{\text{Earnings of aged} + \text{Money value of unpaid family labour} - \text{Consumption expenditure on aged}}{\text{Household expenditure}}$$

Mean net contribution is negative (−Rs.8962). A substantial gender gap is noted in all cases—*households spend more on male respondents, who make a higher gross contribution but lower net contribution.*

We have also estimated net contribution as a percentage of monthly per capita household expenditure. As aged consume more than what they earn plus their imputed contribution to family, the mean is negative. *Net contribution of male respondents (−19.40) is lower than that of females (−14.47).* This is statistically significant. In fact, *only 17.99% of respondents make a positive net contribution to their families.* This contradicts the unrealistically high perception-based results reported in an UNFPA study in India (UNFPA 2017).

Econometric analysis reveals that that *there is no gender difference in absolute net contribution; but, in percent terms, males contribute less than female respondents.* The first OLS model (for absolute net contribution) indicates that old-old group (aged above 80 years), and those with above HS education contribute less. Lower levels of net contribution are associated with (log) MPCE. Respondents residing in Bhubaneswar also tend to contribute more. In the second model (for net contribution in percent terms) we find that net contribution is higher in larger families, in Bhubaneswar and Ranchi.

7.2.6 Time-Use Pattern of the Aged

The study of how people spend their time constitutes a major theme of research in social science disciplines like anthropology, sociology, economics, and social psychology. In recent years, Time Use Diaries (TUD) have emerged as a reliable and accurate data collection instrument to obtain information on the activity patterns of large populations. A Time Use Diary (TUD) method is a sequential and comprehensive record of the activities undertaken by the persons of interest over the course of a day or days. It is generally recorded by the respondent herself/himself, ideally recorded during or very soon after the time of activity. The diary is divided into regular intervals, during which the participant records what activities they were undertaking for each time interval.

Analysis of the mean time spent on different activities indicates that personal activities, of which sleeping is an important component, comprises the dominant activity for the aged; they are also engaged in SNA and extended SNA activities for a considerable part of the day. A gendered division may be noted here—with *men engaged in economic activities, and women participating in extended SNA activities (even here, male respondents spend more time than women on outside home extended SNA activities)*. Similarly, women respondents are engaged more in personal activities, while men tend to socialize more significantly. In all, *men tend to age more actively than women*.

We have also analysed the proportion of waking time spent on major activity heads. Predictably, male aged respondents spend a greater time in economic activities, compared to female respondents. In case of extended SNA activities, the opposite is observed.[1] What is interesting is that *if we compare the aggregate time allocated on economic and extended SNA activities together, the total comes to 22.7% and 21.3% for male and female respondents, respectively. The marginal difference[2] indicates that women contribute an equal number of hours to working; however, the fact that these hours are spent on household activities implies that females remain undervalued and under-recognized contributors to family welfare.*

7.2.7 Active Ageing in India

Results also reveal that elderly men tend to socialize significantly more than their female respondents.[3] In all, *respondents spend about one-third of their waking time actively, with male respondents being significantly more active.*[4] In the absence of

---

[1] Both differences are statistically significant: $t = 5.2672$ and $= -4.4728$, respectively for economic activities and extended-SNA activities, with $p = 0.0$ in both cases.
[2] The difference is also statistically insignificant, as $t = 0.7939$ ($p = 0.4276$).
[3] The difference is statistically significant, with $t = 1.71389$ ($p = 0.0436$).
[4] Given $t = 1.9538$ ($p = 0.0256$) the difference is again statistically significant.

alternative estimates it is not possible to conclude whether India is ageing actively; however, one-third appears somewhat on the lower side.

Econometric analysis confirms gender disparity in active ageing: women respondents spend less time actively, compared to their male counterparts. Similarly, as a person ages, she/he becomes less active. Neither religion nor caste appears to have any effect. Interestingly, it is the affluent respondents who are ageing actively. This would imply that it is economic pressures that force the less affluent aged respondents to work hard. However, Table 6.3 shows that this is not necessarily true. While the low-middle income group does work more, so does the high middle-income group. Moreover, the poorest group spends almost one-fifth of their waking times socializing. So it is not economic pressures, but possibly the fact that poorer households live in smaller accommodation and are more closely knit than richer families. Active ageing does not vary across education levels; we find that only those with graduate, postgraduate, professional, or higher levels of education age less actively than less educated respondents. Finally, residents of Ranchi age more actively than residents of Kolkata and Bhubaneswar.

7.3 Asset or Burden?

Our analysis reveals that, *in an economic sense, aged are a burden, or tax, on society*. However, we should keep in mind two points in this context. First, the substantial economic contribution made by currently aged persons in their middle ages and, in particular, *the investment made on their children; this represents a debt of the younger generation to the aged, which is repaid when the younger generation joins the labour force, replacing the superannuated aged*. Second, the *aged also makes a cultural contribution* as they pass on the culture and traditions of their society to the younger generations. There is no way to value this very important service. *If we could add these two components to net contribution, it is more than likely that we would find that aged are making a positive contribution to society*.

7.4 Some Policy Issues

One of the major aspirations of every human being is to live an active and healthy life. This has become a feasible possibility now in many societies due to the medical advancement increasing longevity.

> While we rejoice in living longer and in better health, and with more financial security, we also query how these aspirations can be sustained, through our own behavioural responses and through public policy and institutional reforms and innovations. The challenge for researchers is therefore to identify strategies that are effective in promoting and sustaining activity, independence and health during older ages, with the help of public policies at the national and the local level, by initiatives from civil society organisations as well as by bottom-up behavioural behaviours (Zaidi et al. 2018:1–2).

One of the ways of attaining this aspiration is *active ageing* (WHO 2002).

This calls for identifying means and strategies to ensure that even *aged people can combine high levels of activity (paid, unpaid, and social), improved health physical and mental status, and a greater degree of autonomy and self-reliance.* We argue that such a policy must be based on three pillars: ensuring healthy non-discriminatory labour market conditions, promoting aged friendly environment and meeting the health challenge of the elderly. Details of how to attain these individual goals are discussed below.

7.4.1 *Promoting Workforce Participation of the Aged*

Paul Baltes (1993) has argued that *'young old' workers (between 60 and 70) have substantial potential for physical and cognitive fitness, retain much of their cognitive capacity, and can develop strategies to cope with the gains and losses of ageing.* This implies that postponement of retirement age and retaining the aged in workforce is an important means of ensuring active ageing. This strategy has two advantages—apart from ensuring active ageing and promoting self-confidence within the worker, it also provides financial security, permitting the aged to satisfy their aspirations. Recognition of these advantages had led initial works on active ageing to focus on promoting workforce participation of the aged. Researchers specifically argued that *public policy must be targeted to reduce structural rigidities obstructing entry of elderly workers into the labour market* (Dhar 2014). It calls for an integrated policy, targeting both push and pull factors constraining the re-employment of aged workers.

For instance, employers should be encouraged to discard the notion that elderly workers have outdated skills and low capability for adaptation and appreciate that the experience, skill, and loyalty of such workers can make them a valuable asset. This realization would motivate employers to modify job specifications and operations, and redesign work to facilitate employment of aged workers (Dhar 2014). In this context, it should be kept in mind that aged workers are easily stressed, particularly as their working conditions are poor (OECD 2006). While regulations to improve working conditions should be introduced, a potential solution to reduce work-related stress is to offer flexible working hours for the elderly workers, or even part-time employment. Although this may lead to coordination problems within the workforce, flexibility in labour market regulations can compensate for this negative effect. If regulations permit lower wages to be paid to elderly workers, this can compensate for higher coordination costs (Casey 2004).[5] Demotion to lower ranked

---

[5] Studies reveal that it is the oldest (and youngest) workers who receive minimum wages (OECD 2014).

7.4 Some Policy Issues

but less exhausting work has also been found to be a viable strategy for re-employing aged workers (Josten and Schalk 2010).

Another challenge before the state is to *increase the substitutability between elder and younger workers*. Although (Goldin and Katz 2007) have shown that older workers are rapidly becoming closer skill substitutes for their younger counterparts, a similar trend is yet to be observed in India. This makes training an important issue. Although older workers are reluctant to enter training programmes, increasing duration of the working career may overcome such reluctance—especially if it increases their flexibility with respect to occupational choice. Retraining, however, is costly and employers themselves are often reluctant to invest in retraining of workers, who will not remain in the workforce for long. Such reluctance may be partially overcome if employers are offered tax incentives for employing elderly workers. The introduction of lifelong learning strategies may also help to spread the costs and duration of retraining over a long time span; this will reduce objections from both the employer and employee side.

Finally, *investment in the health of the elderly workers is an important challenge* before policymakers. Poor health is a major reason for withdrawal of elderly from the labour market (Currie and Madrian 1999). On the other hand, aged from low-income households may be forced to work aggravating their poor health. An important challenge before the state is to ensure healthy ageing through a public health policy catering to the needs of the elderly. This will increase the motivation of the aged to work, reducing the dependence on the state social support system.

At the same time, we should sound a note of caution. The experience of the European Union in promoting workforce participation of the aged has shown that it is possible to postpone retirement successfully through appropriate policy measures. But it also reveals possible dangers of such policy measures. *Encouraging the aged to work may create age imbalances in the labour force.* Too few younger cohorts may reduce labour productivity and growth and increase wage costs (Fig. 7.1). This will reduce competitiveness of the Indian economy in the highly

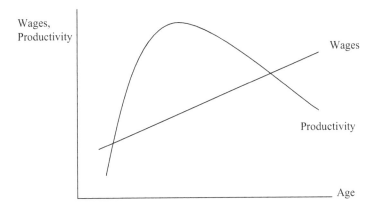

Fig. 7.1 Stylized graph of age, labour productivity, and age. (Source: Storrie 2012)

competitive global market. In the public sector, the fiscal sustainability of pension schemes, already an important issue in India, may be threatened. This implies that there are limits to which measures promoting the aged to work, and we must rely on the other pillars of active ageing.

7.4.2 Other Means of Promoting Active Ageing

In European countries, employment is the central pillar on which active ageing policies are based. This was partly owing to initial researchers focusing on enhanced labour market participation of the aged as the means to ensure on active ageing (Guillemard and Argoud 2004; van den Heuvel et al. 2006). The limited job creation opportunities in the Indian labour market mean that we have to base our strategy on other pillars.

Following the discussions on active ageing at the 1997 Denver Summit (US Department of Health and Human Services 1997), the focus has broadened to include both economically and socially productive activities (Castolo et al. 2004; McKenna 2008). Subsequently, the concept of active ageing was broadened even further to include leisure activities. For instance, Houben et al. (2004) consider activities that require physical and/or mental effort and that occur largely outdoors (social activities). It is on this area that we must focus in tackling the challenge of active ageing.

Boudiny (2013) argues that *age discrimination is a major factor restraining the entrance of the aged in these domains.* Researchers and policy makers have portrayed older people as 'passive, acquiescent, family oriented and disinterested in social and political participation' (Walker and Maltby 2012: 2). We must discard such stereotypes and move towards creating an ageing-friendly environment that ensures optimal participation of the aged in society.

Such age-friendly environment may be created as follows:

> Actions to create age-friendly environments can target different contexts (the home or community, for example) or specific environmental factors (such as transport, housing, social protection, streets and parks, social facilities, health and long-term care, social attitudes and values), and they can be influenced at different levels of government (national, regional or local). When actions also take into consideration social exclusion and barriers to opportunity, these efforts to build and maintain functional ability can also serve to overcome inequities between groups of older adults (WHO 2007:10).

At the same time, policies to foster autonomy of the aged are also important. This is because:

> However, while population-level interventions such as accessible transportation may provide a resource for all older people, some will not be able to benefit fully without individually tailored supports that foster their autonomy and engagement. For example an older woman's ability to be mobile may be determined by her desire to get out and about, and the

7.4 Some Policy Issues

availability of specific mobility devices which correlate to her need (walker, wheelchair, etc.), as well as the level of accessibility and safety of footpaths, buildings, lighting, and the kindness of the bus driver or other passengers to help her get on or off the bus (WHO 2007:10).

Autonomy, therefore, is a core component of the well-being of the aged. It is shaped by their capacity, the spaces they inhabit, their personal resources (such as kinship relations across generations, links with friends, neighbours, and broader social networks) and financial resources they can draw on, and the opportunities available to them. Autonomy depends upon the older person's basic needs being met and on access to a range of services, such as transport, health services, etc. throughout his/her life.

Autonomy can be attained though means like advanced care planning, supported decision-making, and access to appropriate assistive devices. These mechanisms can enable older people to retain the maximum level of control over their lives. Other actions that impact directly on older peoples' autonomy include protecting and ensuring their human rights through awareness-raising, legislation, and mechanisms to address breaches of these rights. In India, legislation has been introduced to prevent the aged from abuse and desertion by their kin. However, access to such protection is limited and skewed across the socio-economic strata.

Another important issue that is still a minor problem, but may become a major future challenge, is that of *orphan elderly* (Margolis and Verdery 2017). This refers to aged residing without spouse, children, or proximate kin; they may reside either alone or in old age homes. While orphan elderly as a proportion of the population is still low (Husain 2018), it is expected to rise because of growing infertility, voluntary childlessness, and marital dissolution. The issue of orphan aged will, therefore, become an important challenge in coming years.

The market is stepping in to provide innovative solutions in the form of OAHs for middle-income and affluent households, and home care (medical or non-medical) services. Such services, however, are costly and can be afforded by only a select few, particularly in urban areas. Further, such agencies need to be regulated to ensure quality, particularly as growing demand may lead to the mushrooming of low-quality and dubious agencies trying to take advantage of the supply deficit. But the other vertices of the care diamond also need to bear responsibility for the orphan aged. The orphan aged has to be made more visible, so that '(t)he medical, public health, and general community … become more aware of these individuals in order to protect and advocate for them' (Carney et al. 2016). Further, public health and community resources have to be mobilized to meet their clearly identified needs, and development of care-giving and decision-making plans has to be given top priority in order to enhance quality of life by optimizing opportunities for health, participation, and security (WHO 2007). Finally, we should also keep in mind that the orphan elderly does not comprise a homogeneous group. They reside alone for different reasons; they live in different circumstances. It is necessary to identify the more vulnerable of them—women, and particularly rural women—and target them specifically.

7.4.3 Health Ageing

Ensuring good health is another important means to active ageing. In this study we have not analysed health status. We suggest that this is an important dimension which needs to be analytically explored in India, particularly the relationship between work and health. Nevertheless, based on existing studies, it is possible to identify the important challenges related to ensuring a healthy aged worker and possible means of how to tackle them.

A recent study (Beard et al. 2015b) points out that the major causes of mortality among aged are ischaemic heart disease, stroke, and chronic obstructive pulmonary disease; further, the burden from these diseases is greater in low-income and middle-income countries vis-à-vis high-income countries. A WHO report on *World Report on Ageing and Health* states that the greatest causes of years living with disability among aged people are sensory impairments, back and neck pain, chronic obstructive respiratory disease, depressive disorders, falls, diabetes, dementia, and osteoarthritis (Beard et al. 2015a). Again, for some, of these disorders, the burden is greater in low-income and middle-income countries.

However, such analysis based on morbidity and mortality patterns reveal only part of health status of the aged. The presence/absence of a health disorder affects satisfaction and perceived health status of the aged. This remains overlooked in morbidity studies. It is also important to take into account the access to health and other services that determine the level of functioning of aged suffering from different disabilities. Multi-morbidity is another issue that adds complexity to the problem of defining research strategies, particularly as more than half of the aged suffer from multi-morbidity (Marengoni et al. 2011; Arokiasamy et al. 2015).

Multi-morbidity has a negative synergistic effect on functioning, quality of life, and mortality risk than individual effects from disorders; this will increase usage of healthcare and increase expenditure on healthcare (Marengoni et al. 2011). Multi-morbidity is a crucial issue for older people in resource-poor settings as it manifests earlier among people living in deprived areas, and people of low socioeconomic status. In developing countries, the problem is aggravated by the double burden of communicable and non-communicable diseases, infections, and the earlier onset of common disorders (Hearps et al. 2014).

The multifaceted dynamics between underlying physiological change, chronic disease, and multi-morbidity can result in health states in older age that are not captured by traditional disease classifications like ICD-10. Notably, one manifestation of the dynamics is the phenomenon of *frailty* (Harttgen et al. 2013), which can severely curtail functioning of the aged. This implies that, instead of building epidemiological profiles of the aged, it is more relevant to study their functioning and the extent to which they are active.

7.4 Some Policy Issues

A WHO report (Beard et al. 2015a) considers the above health issues, underlying physiological and psychosocial changes associated with ageing as interacting to determine an *older person's intrinsic capacity*. 'This capacity is defined as the composite of all the physical and mental (including psychosocial) capacities that an individual can draw on at any point in time' (Beard et al. 2015a, b:5). An examination of intrinsic capacity over life cycle undertaken by the WHO report reveals

> a gradual decline in intrinsic capacity with increasing age across the life course. Of course, for individuals, any decrease with increasing age might not be smooth, but instead consist of intermittent setbacks and recoveries. But for the population as a whole, the average reduction shown in this analysis was gradual. There was no age when people suddenly had less capacity and became "old" (Beard et al. 2015b).

The level of intrinsic capacity to function also depends upon socio-economic status. This implies that it is typically people who are poor and have least access to health services are most in need of support to improve their functioning. The physical space or environment—comprising of housing, transport, access to different services—in which the individual resides, and their interaction with this environment, will determine an individual's intrinsic *ability* to undertake activities important to him or her, transcending the limits imposed by his/her intrinsic *capacity* to perform certain tasks.

The WHO report identifies *four priority areas for action that can help to ensure an optimal (or near optimal) life trajectory of intrinsic functioning*, based on these two concepts of capacity and ability (Beard et al. 2015a):

1. Health systems need to be integrated and aligned to the older populations they serve;
2. Systems should be developed to provide long-term care to maintain a level of functional ability in older people who have, or are at high risk of, substantial losses of capacity, and to ensure this care and support is consistent with their basic rights, fundamental freedoms, and human dignity.
3. An age-friendly environment should be promoted through creation and maintenance of intrinsic capacity (either through reduction of health risks, encouragement of capacity-enhancing behaviours or removal of barriers to them, or by provision of services that foster capacity), and enable greater functional ability in someone with a particular level of capacity to ensure that aged can meet their basic needs; learn, grow, and make decisions; move around; build and maintain relationships; and contribute.
4. Progress on healthy ageing can be attained only by filling in currently existing knowledge and research gaps. These gaps include an absence of consensus about how to define, measure, and analyse key concepts; the exclusion of older people from many population surveys and even from clinical trials for treatments for which they will be the major recipients; and economic analyses that fail to consider the contributions that older people make to society.

References

Arokiasamy, Perianayagam, Uttamacharya Uttamacharya, Kshipra Jain, Richard Berko Biritwum, Alfred Edwin Yawson, Wu Fan, Yanfei Guo, et al. 2015. The impact of multimorbidity on adult physical and mental health in low- and middle-income countries: What does the study on global ageing and adult health (SAGE) reveal? *BMC Medicine* 13: 178. https://doi.org/10.1186/s12916-015-0402-8.

Baltes, P.B. 1993. The aging mind: Potential and limits. *The Gerontologist* 33: 580–594.

Beard, John, Alana Officer, and Andrew Cassels. 2015a. *World report on ageing and health*. Luxembourg.

Beard, John R., Alana Officer, Islene Araujo de Carvalho, Ritu Sadana, Anne Margriet Pot, Jean-Pierre Michel, Peter Lloyd-Sherlock, et al. 2015b. The World report on ageing and health: A policy framework for healthy ageing. *The Lancet* 387: 2145–2154. https://doi.org/10.1016/S0140-6736(15)00516-4.

Boudiny, Kim. 2013. "Active ageing": From empty rhetoric to effective policy tool. *Ageing and Society* 33. Cambridge University Press: 1077–1098. https://doi.org/10.1017/S0144686X1200030X.

Carney, Maria T., Janice Fujiwara, Brian E. Emmert, Tara A. Liberman, and Barbara Paris. 2016. Elder orphans hiding in plain sight: A growing vulnerable population. *Current Gerontology and Geriatrics Research* 2016. Hindawi: 1–11. https://doi.org/10.1155/2016/4723250.

Casey, Bernard. 2004. The OECD jobs strategy and the European employment strategy: Two views of the labour market and the welfare state. *European Journal of Industrial Relations* 10: 329–352. https://doi.org/10.1177/0959680104047024.

Castolo, Octavio, Filipa Ferrada, and Luis Camarinha-Matos. 2004. Telecare time bank: A virtual community for elderly care supported by mobile agents. *The Journal on Information Technology in Healthcare* 2: 119–133.

Currie, Janet, and Brigitte C. Madrian. 1999. Health, health insurance and the labor market. In *Handbook of labor economics*, vol. 3, 3309–3416. Amsterdam: North Holland: Elsevier. https://doi.org/10.1016/S1573-4463(99)30041-9.

Dhar, Antara. 2014. Workforce participation among the elderly in India: Struggling for economic security. *The Indian Journal of Labour Economics* 57: 221–245.

Goldin, Claudia, and Lawrence Katz. 2007. *Long-run changes in the U.S. wage structure: Narrowing, widening, polarizing*. Vol. 13568. Cambridge, MA. https://doi.org/10.3386/w13568.

Guillemard, A.-M., and D. Argoud. 2004. France: A country with a deep early exit culture. In *Ageing and the transition to retirement: A comparative analysis of European welfare states*, ed. T. Maltby, B. de Vroom, M.L. Mirabile, and E. Øverbye, 165–185. Aldershot: Ashgate.

Harttgen, Kenneth, Paul Kowal, Holger Strulik, Somnath Chatterji, and Sebastian Vollmer. 2013. Patterns of frailty in older adults: Comparing results from higher and lower income countries using the Survey of Health, Ageing and Retirement in Europe (SHARE) and the study on global AGEing and adult health (SAGE). Edited by Jerson Laks. *PLoS ONE* 8.. Public Library of Science: e75847. https://doi.org/10.1371/journal.pone.0075847.

Hearps, Anna C., Genevieve E. Martin, Reena Rajasuriar, and Suzanne M. Crowe. 2014. Inflammatory co-morbidities in HIV+ individuals: Learning lessons from healthy ageing. *Current HIV/AIDS Reports* 11: 20–34. https://doi.org/10.1007/s11904-013-0190-8.

Houben, M., V. Audenaert, and D. Mortelmans. 2004. Vrije tijd en tijdsbesteding [Leisure and time allocation]. In *Op latere leeftijd. De leefsituatie van 55-plussers in Vlaanderen*, ed. T. Jacobs, L. Vanderleyden, and L. Boer Vanden, 329–342. Antwerp: Garant.

Husain, Zakir. 2018. *The long day closes: The orphaned elderly in post-globalised India*, Working paper. Kharagpur: Humanities & Social Sciences Department, Indian Institute of Technology.

Josten, Edith, and René Schalk. 2010. The effects of demotion on older and younger employees. *Personnel Review* 39: 195–209. https://doi.org/10.1108/00483481011017417. Emerald Group Publishing Limited.

References

Marengoni, Alessandra, Sara Angleman, René Melis, Francesca Mangialasche, Anita Karp, Annika Garmen, Bettina Meinow, and Laura Fratiglioni. 2011. Aging with multimorbidity: A systematic review of the literature. *Ageing Research Reviews* 10: 430–439. https://doi.org/10.1016/j.arr.2011.03.003.

Margolis, Rachel, and Ashton M. Verdery. 2017. Older adults without close kin in the United States. *Journals of Gerontology – Series B Psychological Sciences and Social Sciences* 72: 688–693. https://doi.org/10.1093/geronb/gbx068.

McKenna, Margaret A. 2008. Transcultural nursing care of older adults. In *Transcultural concepts in nursing care*, ed. Margaret M. Andrews, 5th ed., 168–194. Philadelphia: Wolters Kluwer Health/Lippincott Williams & Wilkins.

OECD. 2006. *Live longer, work longer. Ageing and employment policies*. Paris: OECD Publishing. https://doi.org/10.1787/9789264035881-en.

———. 2014. *Ageing and employment policies: Netherlands: Working better with age*, Ageing and employment policies. Netherlands: OECD Publishing. https://doi.org/10.1787/9789264208155-en.

Storrie, Donald. 2012. Living longer – Working better. In *Thematic seminar on employment policies to promote active ageing by employment*. Brissels: Social Affairs and Inclusion, European Commission.

UNFPA. 2017. *Caring for our elders: Early responses caring for our elders: Early responses*. New Delhi.

US Department of Health and Human Services. 1997. *Active aging: A shift in the paradigm*. Denver.

van den Heuvel, N., W. Herremans, P. van der Hallen, C. Erhel, and P. Courtioux. 2006. De arbeidsmarkt in Vlaanderen [The Labour Market in Flanders]. In *Special issue: Active ageing, early retirement and employability*. Antwerp: Garant.

Walker, Alan, and Tony Maltby. 2012. Active ageing: A strategic policy solution to demographic ageing in the European Union. *International Journal of Social Welfare* 21: S117–S130. https://doi.org/10.1111/j.1468-2397.2012.00871.x. Wiley/Blackwell (10.1111).

WHO. 2002. *Active ageing: A policy framework*. Geneva.

———. 2007. *Women, ageing and health: A framework for action*. Geneva.

Zaidi, Asghar, Sarah Harper, Kenneth Howse, Giovanni Lamura, and Jolanta Perek-Białas. 2018. Towards an evidence-based active ageing strategy. In *Building evidence for active ageing index and its potential*, 1–15. Singapore: Plagrave, Macmillan. https://doi.org/10.1007/978-981-10-6017-5_1.

CPSIA information can be obtained
at www.ICGtesting.com
Printed in the USA
LVHW080852011220
673037LV00001B/54